Thermal Insulation

Thermal Insulation

Edited by

S. D. PROBERT

SCHOOL OF ENGINEERING,
UNIVERSITY COLLEGE,
SWANSEA

and

D. R. HUB

ORGANIZER OF SHORT COURSES,
INSTITUTE OF SCIENCE AND TECHNOLOGY,
UNIVERSITY OF WALES,
CARDIFF

ELSEVIER PUBLISHING CO LTD
AMSTERDAM – LONDON – NEW YORK
1968

ELSEVIER PUBLISHING CO. LTD.
BARKING, ESSEX, ENGLAND

ELSEVIER PUBLISHING CO.
335 JAN VAN GALENSTRAAT
P.O. BOX 211 AMSTERDAM

AMERICAN ELSEVIER PUBLISHING COMPANY INC.
52 VANDERBILT AVENUE
NEW YORK, N.Y.

© ELSEVIER PUBLISHING CO. LTD., 1968

LIBRARY OF CONGRESS CATALOG CARD NUMBER 68-17578

Printed in Great Britain by Galliard Limited, Great Yarmouth, England

Contents

	Page
Introduction	vii

Low-Temperature Insulation
 A. H. Cockett, Research and Development Department, British Oxygen Co. Ltd., Great Britain. — 1

Superinsulation in Cryogenic Engineering
 S. Mercer, Cryosystems Ltd., Great Britain. — 15

A Mechanically Strong Thermal Insulator for Cryogenic Systems
 S. D. Probert and T. R. Thomas, School of Engineering, University College, Swansea; and D. Warman, Department of Applied Physics, Institute of Science and Technology, Cardiff — 29

Low-Temperature Applications of Expanded Polystyrene and Expanded Ebonite
 H. S. F. Baker, Onazote Insulation Company, London, Great Britain. — 38

Applications of Rigid Polyurethane Foam
 W. J. Wilson, Dyestuffs Division, Imperial Chemical Industries Ltd., Manchester, Great Britain. — 45

Thermal Insulation using Multiple Glazing
 Thomas A. Markus, Professor of Building Science, University of Strathclyde, Great Britain. — 62

Dual Purpose Materials: Thermal Insulation and Sound Absorption
 J. Lawrie, Structural Insulation Association, London, Great Britain. — 83

The Design of Walls for Intermittent Furnaces Using High Temperature Insulation

    W. H. Holmes, The British Ceramic Research Association, Great Britain.   94

Selection and Comparison of Materials Using Figures of Merit

    P. H. H. Bishop, C.P.M. Department, Royal Aircraft Establishment, Farnborough, Great Britain.   110

Index   119

Introduction

The minimisation of energy wastage is as important as finding new sources of fuel. Therefore, as more commercial processes are undertaken at elevated or reduced temperatures, the prevention of unwanted heat leaks has received increasing attention, not only to reduce running costs but also to simplify the stabilisation of temperatures over large volumes—a common industrial requirement. Impetus for improved thermal insulation has been given by the demand for better power utilisation, by the greater comfort obtainable (in both summer and winter) and by the prevention of condensation. The last-named corrodes metalwork, spoils decorations, rots timbers and increases heat losses and general dampness. Recently, thermal insulation research has been fostered by the space race (for which rockets fuelled with liquid hydrogen and liquid oxygen boiling at $-253°C$ and $-183°C$, respectively, are used) and by nuclear engineering projects involving ultra-high temperatures. Such insulation studies have resulted in a considerable "fall-out" of technical information, some of which has been successfully employed in domestic engineering.

A symposium outlining work in thermal insulation was held at the Welsh College of Advanced Technology, Cardiff, from 4–6th October, 1965. (Historically it was appropriate that this first thermal insulation symposium should occur in the Principality, since the first mineral wool, used mainly for insulating boilers and steam pipes, was produced there in 1840.) An amplified and up-to-date version of selected lectures reviewing some aspects of the subject has been collated in the following pages. From these it can be seen that thermal insulation has almost won the struggle to rise above its empirical foundations and so become an applied science. Nevertheless, too often thermal insulation is still an afterthought, whereas it should be included at the design stage. Builders, for example, rarely provide adequate insulation for the roof or double glazing for north-facing windows. However, the return on the capital invested in properly designed and installed insulation can easily exceed 20% per annum.

With insulation the cost of the *completed* system is of primary importance, and not merely the cost of materials. The shape of the system often dictates the insulant to be used (see Figs. 1 and 2). Powders may be chosen to increase the thermal resistance of a spherical cavity, as they can be poured in and therefore do not involve high labour charges. However, in all cases, consideration of thermal insulation should be given at the *design* stage, because "afterthought" insulation is invariably more expensive than if the appropriate steps had been taken upon initial assembly. Also there is little thermal gain by having extra wall-cladding if the design does not include appropriate measures to inhibit convective losses from within the "protected" environment.

Engineers and architects have available numerous materials with different combinations of heat insulation and thermal inertia. The choice of insulant for a particular application is inevitably a compromise dependent upon cost and other factors, which may include:

(i) whether reduction of the rate of heat exchange is more important than the enclosed environment rapidly attaining a uniform temperature;

Fig. 1. Application of fibreglass rigid section used to insulate steam pipes in a modern car factory. (Reproduced by kind permission of Fibreglass Ltd.)

(ii) the extreme surface temperatures to which it is subjected;
(iii) the permitted mechanical and thermal stresses, so that distortion and shrinkage of the insulation do not occur which result in thermal transmission windows;
(iv) rigidity;

Fig. 2. Fibreglass insulation in place.

(v) resistance to damage and deterioration in service due to the presence of vibration and dampness;
(vi) density;
(vii) external finish;
(viii) flammability; and
(ix) corrosivity, odour and health hazards during installation.

ERRATUM

On page ix Fig. 2 has unfortunately been printed upside down.

Some of these factors will be considered in detail in the following chapters.

Simple criteria for deriving the optimum thickness of insulation are given in British Standard 3708. Nevertheless, the calculations are often more complex than this suggests. For example, in a power station an appreciable part of the heat escaping from the boiler casings and pipework is returned to the system by the boiler fans. In more refined analyses, the effects of thermal bridging *via* supports such as pipe hangers also need to be considered. Engineers and designers are urged to consult their latest national standard.

Considerable avoidance of idleness of site labour is possible with the advent of industrialised building. This involves almost completely dry construction and prefabrication as well as prefinishing of components in the factory, all of which simplify the introduction of thermal insulation. This is mainly applied in the factory, though there are still instances where application on site is necessary, *e.g.*, at the edges of floors and walls where these meet the cladding. Industrialised building also facilitates better design—soon it is likely that the four exterior walls which are subjected to very different thermal assaults (*e.g.*, wind, thermal radiation, etc.) will not necessarily be built of the same materials and to the same thicknesses. This is standard practice for most buildings at the present time.

1

Low-temperature Insulation

A. H. COCKETT
Research and Development Department, British Oxygen Co. Ltd. (Great Britain)

INTRODUCTION

This chapter deals with the problem of keeping atmospheric heat out of vessels, plant or buildings, of which the contents have to be maintained at temperatures below about 100°K (−173°C). At this temperature the insulant can have gas- (*i.e.*, air-) filled pores, but below about 78°K air liquefies, and therefore the pores must be evacuated. Because of the difference in the fundamental heat transfer processes involved it is convenient to consider separately the two different classes of insulants.

Low-temperature insulation techniques make use of some of the properties of materials peculiar to the temperature range concerned and these properties are therefore briefly reviewed in an introductory section.

LOW-TEMPERATURE PHYSICAL PROPERTIES

Thermal conductivity

Pure metals

Heat conduction in pure metals is largely electronic, thermal resistance being due to scattering by impurities, defects and the crystal lattice. At the lowest temperatures scattering by impurities and defects predominates and the thermal resistance is inversely proportional to the temperature. At higher temperatures lattice scattering is the more important, and resistance is proportional to the square of the absolute temperature, *i.e.*, T^2. At any temperature the total resistance is the sum of those two, and hence has a minimum at some intermediate temperature, typically 10–20°K, and the conductivity shows a maximum at this temperature. The height of the peak of the thermal conductivity–temperature curve depends on the purity of the metal.

Alloys

In alloys lattice vibrations contribute most of the conductivity, the resistance being inversely proportional to T^2. With some electronic conduction the total conductivity is of the form $K = \alpha T^2 + \beta T$. However, for industrially important alloys no simple theory applies, but the conductivity decreases with temperature.

Dielectric crystals

In dielectric crystals, heat is conducted by lattice vibrations, and at the lowest temperatures is limited by boundary reflection, giving a thermal conductivity proportional to T^3, and varying with specimen size. At higher temperatures, vibrations of different wavelengths interfere to give a conductivity proportional to $e^{-\theta/gT}$. Thus dielectric crystals, like pure metals, have a conductivity maximum at low temperatures.

Gases

Kinetic theory indicates that conductivity is proportional to $\sqrt{T/M}$ for ideal gases, and for real gases experiment gives a similar relation, $k \alpha T^n$, where n is approximately 0·5. Sutherland's empirical equation $k = \dfrac{aT^{\frac{1}{2}}}{1 + \dfrac{C}{T}}$ is also of wide application and gives closer agreement with observation.

Numerical values

In much of the literature thermal conductivity is expressed in watts/cm deg C, and here the same units will be used. For the convenience of those more familiar with other units the following conversion factors are given.

A thermal conductivity of 1 watt/cm deg C = 0·239 cal/cm s deg C
= 57·8 Btu/ft h deg F
= 692 Btu in/ft² h deg F

A heat transfer coefficient of
1 watt/cm² deg C = 0·239 cal/cm² s deg C
= 1761 Btu/ft² h deg F.

The wide range of thermal conductivity exhibited by commonly available materials is shown in Fig. 1.1, X indicating the typical value for insulants, with more data for these appearing in Table 1.

Insulants

The conductivities of many solids used as insulants are about 0·008 watt/cm deg C at ambient temperature while that of air is about ·00015 watt/cm deg C. The common insulants are dispersions of these solids in air, the best of them having thermal conductivities not much greater than that of air. Table 1 shows for some examples the mean thermal conductivity over the temperature range 90–300°K.

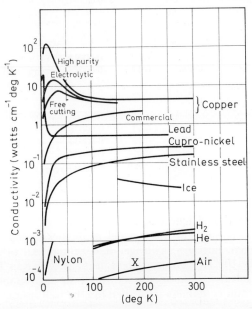

Fig. 1.1. Thermal conductivity at low temperatures.

Other desirable attributes of insulants

In addition to low thermal conductivity, insulants must have other important qualities. They should, for instance, be:

(a) Non-flammable even in presence of oxygen; this limits the choice to inorganic materials for oxygen itself, but not for helium or hydrogen.

(b) Non-toxic.

(c) Non-corrosive, even in presence of moisture, to materials used in constructing plant and containers, including mild steel, stainless steel, aluminium and copper.

(d) Dimensionally stable, particularly when used on mobile equipment.

(e) Cheap.

THERMAL INSULATION

TABLE 1
SOME PROPERTIES OF INSULANTS

	Solid			Insulant		
Material	Density (g/cm^3)	Conductivity (watts/cm deg K at 300°K)	Density	Mean conductivity (watts/cm deg K, 90–300°K)	Vol. % of solid	Material
Air				0·00015		
Glass	2·5	0·008	0·19	0·00025	7·6	Mineral wool
Silica	2·21	0·0096	0·08	0·00015	3·6	Santocel
Basic magnesium carbonate	1·7		0·13	0·00026	7·6	BMC powder
PVC	1·5	0·0017	0·12	0·00034	8·0	Expanded PVC
Ebonite	1·15		0·064	0·00027	5·6	Onazote
Polystyrene	1·06	0·0010	0·016	0·00022	1·5	Expanded polystyrene

Thermal properties of liquefied gases

Now consider some of the thermal properties of liquefied gases such as oxygen, nitrogen, hydrogen and helium. All these liquids, boiling below 100°K, have low latent heats as shown in Table 2. A small heat inleak would inevitably result in appreciable loss by evaporation and containers must therefore be well insulated.

TABLE 2
SOME PROPERTIES OF LIQUEFIED GASES AT THEIR NORMAL BOILING POINT

Liquid	Boiling point (°K)	Density (g/cm^3)	Latent heat (J/g)	Volumetric latent heat (kJ/l)	Evaporation rate for 1 watt (cm^3/h)
O_2	90·2	1·140	213	242	14·7
N_2	77·3	0·808	199	161	22·2
Ne	27·3	1·204	88	106	34·1
H_2	20·4	0·071	455	32·2	111
He	4·2	0·1248	20·8	2·60	1380

Conversion factors: 1 J/g = 0·239 cal/g
= 0·43 Btu/lb
1 kW = 3412 Btu/h
1 W = 3·41 Btu/h.

Because of the low boiling point much power has to be expended to recondense the vapour produced by any heat inleak, and the cost of this expensive process can be diminished by improving the insulation. The relative incentives in the case of the liquids under consideration can be estimated by considering a refrigerator, which is a heat engine in reverse, removing heat at the low boiling point T_2 and rejecting it at ambient temperature T_1. Assuming that it operates on the Carnot cycle the coefficient of performance, *i.e.*,

$$\frac{\text{Refrigerating effect}}{\text{Work performed}} = \frac{T_2}{T_1 - T_2}.$$

For a practical cycle and imperfect machine with overall efficiency η,

the coefficient of performance $= \dfrac{\eta T_2}{T_1 - T_2}.$

Taking T_1 as 300°K, the figures shown in Table 3 have been calculated for the liquids under discussion, the last column showing the rate of power dissipation to recondense vapour equivalent to one litre of liquid per hour.

TABLE 3
POWER REQUIRED TO RECONDENSE COLD VAPOUR

Liquids	T_2 (°K)	$\dfrac{T_1 - T_2}{T_2}$	η	$\dfrac{T_1 - T_2}{\eta T_2}$	Watts at 300°K to recondense 1 litre/hour
CH_4	111·7	1·68	0·25	7	400
O_2	90·2	2·33	0·25	9	600
N_2	77·3	2·94	0·23	13	600
Ne	27·3	9·96	0·09	111	3000
H_2	20·4	13·67	0·07	195	2000
He	4·2	70·5	0·04	1800	2000

Tables 2 and 3 both indicate a marked incentive to improve the insulation of liquids boiling below 70°K, the same point of demarcation as mentioned previously. Above this temperature conventional insulants such as slag wool and expanded perlite, used at higher temperatures, are generally satisfactory. Below it materials of much lower thermal conductivity are required.

THERMAL INSULATION

CONVENTIONAL INSULANTS FOR LIQUEFIED ATMOSPHERIC GASES

Liquefied atmospheric gases are processed and produced in low-temperature distillation columns. As objects to be insulated these may be regarded as columns typically 4 ft in diameter and 30 ft tall containing a mixture of cold liquids and vapour, the temperature being 81°K ($-192°$C) at the top and 100°K ($-173°$C) at the bottom as shown in Fig. 1.2. A number of pipes provide connections between

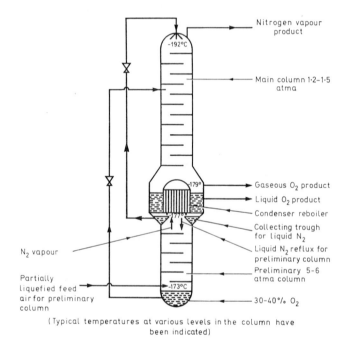

(Typical temperatures at various levels in the column have been indicated)

Fig. 1.2. Linde double column for air separation (as at foot of Figure).

different levels, and subsidiary columns may be attached for separating special components such as the rare gases. These make for difficulty in packing the insulant evenly. Because of the very low temperature, the surrounding air in contact with the column attains a very high density, as this is inversely proportional to T. There is therefore a strong tendency towards convection, for which the driving force may be seven times as great as can be achieved at any temperature above ambient. Hence, insulants of low air permeability are required, and slag wool is commonly used. Air permeability diminishes with increasing packing density, but thermal conductivity varies as shown for ambient and low temperatures in Fig. 1.3. Thus the optimum packing density of 8–12 lb/ft^3 is used. Expanded perlite has also been used, introduced into the outer casing pneumatically. It flows readily into all the interstices but must be completely removed for maintenance. The base of the column is supported on timber or foamed glass bricks, with slag wool insulation, or on concrete pillars providing an air space above the ground to prevent frost heave.

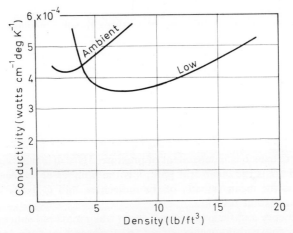

Fig. 1.3. Variation of thermal conductivity of slag wool with packing density at ambient and low temperatures.

Pipelines in many air separation plants are insulated by slag wool in an outer metal casing 9–12 inches in diameter. Preformed expanded PVC can be used for pipes conveying liquid nitrogen and normal boilerhouse pipe insulants can be used on pipelines intended for intermittent use.

For liquid methane, in pipes up to 14 in diameter, 6 in of expanded ebonite has been used as insulant applied in three layers each 2 in thick.

For many years liquid oxygen and liquid nitrogen tanks were spherical in shape with an inner container supported by chains in a larger vessel, the interspace being filled with 12 in of basic magnesium carbonate, a dense powder permitting no convection. The outer casing was sealed. Many modern tanks have the interspace filled with expanded perlite. Preformed insulant blocks shaped to the tank contour have been found unsatisfactory, on account of convection in the space between them. Moulded tanks of foamed insulant have been tried, but these failed because of the cracks produced by thermal contraction. The cracks start on the inner surface but eventually penetrate completely through the insulant. Most recent tanks have improved insulants such as are described later.

DEVELOPMENTS IN LOW TEMPERATURE INSULATION

Heat inleak can occur by conduction through solid insulant and gas filling the pores, by convection in the gas, and by radiation. Conduction through the solid having been minimised it is necessary next to consider heat transfer through the gas to find a means of reducing the overall heat inleak.

Heat transfer through the gas

For an ideal gas the conductivity $k = \frac{1}{3}m\lambda v \bar{c} C_v$ and, since $\lambda \propto \frac{1}{v}$, $k \propto mcC_v$ which is independent of pressure. Here m is the weight of a molecule, λ is the mean free path, v is the number of molecules per cm^3, $\bar{c}$ is the mean velocity of the molecules, and C_v is the specific heat at constant pressure.

The theory and formulae break down when mass movement of the gas (i.e., convection) occurs, giving an increased heat flow, and when at low pressure or in small spaces the mean free path λ is comparable with the dimensions. Knudsen showed that in the latter case the heat transferred per unit area between parallel plates $= H = A\frac{1}{2}(\gamma + 1) \frac{pC_v}{\sqrt{2\pi RT}}$ where $A = \frac{a_1 a_2}{a_1 + a_2 - a_1 a_2}$, a_1 and a_2 being the areas of the plates and γ the specific heat ratio. The heat flow H is independent of the separation (except that this must be greater than λ).

The curve marked "void" in Fig. 1.4 shows the heat transfer to a 5-litre spherical copper vessel containing liquid oxygen at 90°K from an outer shell at 300°K at various gas pressures in the interspace. At high pressures the heat inleak is high because of convection; at intermediate pressures it is independent of pressure in agreement with kinetic theory, and at low pressures it diminishes with pressure.

With a powder or other disperse solid in the same shell, similar results would be expected, except that since the spaces are smaller, higher pressures can be tolerated. This is confirmed by experiment as indicated in Fig. 1.4, which gives results for a number of insulants in the space between the shells of the same 5-l vessel. Mean thermal conductivities over the temperature range 90–300°K measured under high vacuum are much lower even than the conductivity of air, typical figures from these measurements being:

$$\begin{array}{ll} \text{Brelite} & 0\cdot000023 \text{ watt/cm deg C} \\ \text{Santocel} & 0\cdot000010 \text{ watt/cm deg C} \\ \text{Microcel E} & 0\cdot000008 \text{ watt/cm deg C.} \end{array}$$

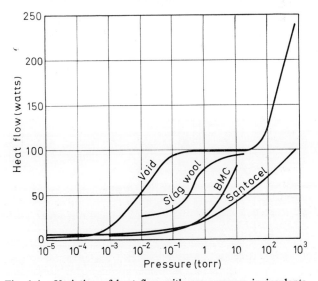

Fig. 1.4. Variation of heat flow with gas pressure in insulants.

In Fig. 1.4 there are small but finite intercepts at zero pressure which indicate some form of heat transfer not due to the gas, *i.e.*, some solid conduction and some radiation. In this case the solid conduction was due to the supports of the inner 5-l vessel, but the total is so small that it is worthwhile considering the effect of radiative heat transfer even though this is also small, at least by comparison with conduction through conventional insulants.

Heat transfer by radiation

The heat radiated by a surface at temperature T is proportional to T^4, and to its emissivity ε. Between two surfaces of the same emissivity and of area A, one at high temperature T_1 and the other at lower temperature T_2, the heat flow is

$$H = A\frac{\varepsilon}{2}\sigma(T_1^4 - T_2^4)$$

where σ is Stefan's constant.

Reducing T_1 by cooling the outer wall is an obvious means of reducing the heat inleak. For hydrogen and helium the outer walls can be cooled by liquid nitrogen, as illustrated in Fig. 1.5. For liquid oxygen or liquid nitrogen there is no convenient refrigerant but it is possible to use cold vapour.

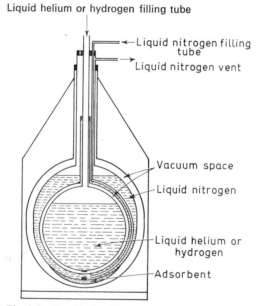

Fig. 1.5. Duplex liquid nitrogen-shielded Dewar.

Another way of reducing heat transfer by radiation is by inserting a multiplicity of thermally insulated screens of low emissivity in the evacuated space. Consider, for simplicity, two infinite plane parallel surfaces of area A at temperatures T_1 and T_2, spaced unit distance apart. In the intervening space insert $N - 1$ screens of emissivity ε and let the N spaces be filled with a transparent conducting medium of conductivity k. The heat transfer from screen r to screen (r + 1) will then be

$$Q = NkA\,(T_r - T_{r+1}) + \frac{\sigma A}{(2/\varepsilon - 1)}\,(T_r^{\,4} - T_{r+1}^{\,4}).$$

Assuming ε and k to be independent of temperature, for the whole system the heat transfer is found, by summing N similar equations, to be

$$Q = Ak\,(T_1 - T_2) + \frac{RA}{N}\,(T_1^{\,4} - T_2^{\,4}),\ \text{where}\ R = \frac{\sigma}{2/\varepsilon - 1}.$$

If this is divided by the surface area and the difference between the temperatures T_1 and T_2 of the inner and outer walls, we obtain

$$k_{12} = \frac{Q}{A(T_1 - T_2)} = k + \frac{R}{N}(T_1 + T_2)(T_1^{\,2} + T_2^{\,2}),$$

which gives the relation between an effective thermal conductivity k_{12}, the properties of the insulant and the temperatures of the walls. Similar equations can be derived for cylindrical and spherical concentric surfaces. Measurement of the heat inleak with a range of values of T_1 makes possible separation of the two components, the thermal conductivity k and the contribution due to radiation.

Multilayer insulants of this type designed to eliminate radiant heat transfer were studied in the vessel shown in Fig. 1.6. This generally resembled a glass Dewar flask with the inner container separated into two sections, the upper one preventing heat from the atmosphere above reaching the lower. Thus all the heat reaching this one had traversed the test insulant surrounding its vertical cylindrical sides and hemispherical base. The heat was calculated by measuring the rate of gas production caused by evaporation of the contained liquid (nitrogen). Some results are given in Table 4 and Fig. 1.7.

The lowest conductivities were obtained with multiple screens of aluminium foil separated by insulating fabric net, and this type of insulant can be applied to containers for liquid helium and liquid hydrogen.

Still lower figures have been obtained with aluminium foil and glass fibre paper, insulants suitable for liquid oxygen and liquid nitrogen. Powder insulants are very convenient and easy to apply, and the table shows that a transparent one like Santocel A for which

96% of the heat transfer is by radiation can be admixed with aluminium powder with great reduction in effective conductivity. With the 40/60 mixture 60% of the much reduced heat transfer is by radiation. Fig. 1.8 shows the way in which the thermal conductivity of such a mixture varies with the proportions of the two powders.

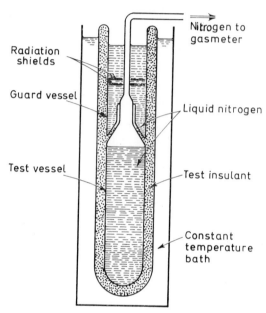

Fig. 1.6. Vessel for measuring thermal conductivity of insulants.

TABLE 4
CONDUCTION AND RADIATION THROUGH SELECTED INSULANTS

Insulant	Effective conductivity k_{12} ($\mu\,watt/cm\,deg\,C$)	Conduction k ($\mu\,watt/cm\,deg\,C$)	Radiation ($\mu\,watt/cm\,deg\,C$)	(Per cent of k_{12})
Santocel A	16·3	0·6	15·7	96
40/60 (w/w) Aluminium Santocel	3·4	1·4	2·0	60
Microcel E	8·3	3·5	4·8	58
Foil-net	1·8	0·6	1·2	66
Copper Dewar	6·2	1·7	4·5	72

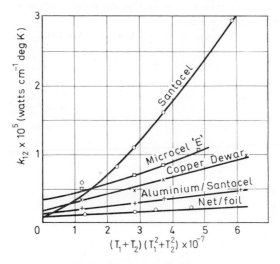

Fig. 1.7. Variation of effective thermal conductivity k_{12} with temperature of warm surface.

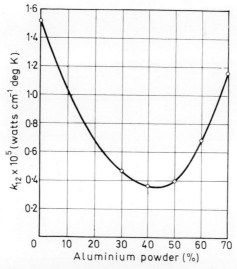

Fig. 1.8. Variation of effective thermal conductivity k_{12} of Santocel–aluminium powder with composition.

Microcel E is not transparent and its conductivity is not reduced by adding aluminium powder.

Insulants like microcel- and santocel–aluminium or santocel–copper mixtures are used for insulating liquid oxygen and liquid nitrogen tanks. The heat inleak and therefore the evaporation rates have been reduced by a factor of 5 or more by comparison with older designs using conventional insulants in air at atmospheric pressure. With the multilayer insulants liquid hydrogen and helium can be stored without resorting to liquid nitrogen cooling of the outer wall.

REFERENCES

1. S. S. Kistler, *J. Phys. Chem.*, **39** (1935) 79.
2. R. H. Kropschot, B. J. Hunter and M. M. Fulk, *Paper C*3, Proc. Cryogenic Eng. Conf., 1959.
3. P. Petersen, *T. V. F.* (*Sweden*), **29** (1958) 151.
4. A. H. Cockett and W. Molnar, *Cryogenics*, **1** (1960) 21.
5. F. Din and A. H. Cockett, *Low Temperature Techniques*, G. Newnes, London, 1960.
6. R. B. Scott, *Cryogenic Engineering*, D. van Nostrand Co. Inc., Princeton, N.J., 1959.

2

Superinsulation in Cryogenic Engineering

S. MERCER
Cryosystems Ltd. (Great Britain)

INTRODUCTION

Insulation is used as a thermal barrier to minimise or reduce the flow of heat between warm and cold boundaries. On the one hand it is used to prevent heat leaving a system and on the other to prevent heat entering a system. In each case the object is the same but the temperature to be preserved is very different.

Whether a system is warm or cold it is usually insulated to reduce the flow of heat to or from its surroundings at ambient temperature. In the first case the system is warmer than its surroundings and in the second case it is colder. If insulation is not used, it will be necessary to supply heat in the first case and to subtract heat in the second.

Heat can be generated in various ways and it is relatively cheap. Refrigeration is more difficult to produce and is more expensive. Figure 2.1 shows the approximate cost of producing refrigeration which rises rapidly with falling temperature. As the temperature of a refrigerant approaches absolute zero the cost of producing it becomes very high, hence the incentive to preserve it is high also.

The insulation of a system can be justified only if the reduction in heat transfer results in an economic gain. In other words, the cost of insulating a system must be less than the saving in cost of heat or refrigeration. Superinsulation is a very expensive form of insulation; its use is economical only to conserve very low temperatures.

Cold or refrigeration is produced and stored in the form of liquefied gases, of which the principal ones are helium, hydrogen and nitrogen. Low-temperature insulants are used primarily to preserve liquefied gases in the liquid state. Some of the physical properties of the more commonly used low-boiling fluids are given in Table 1 as are the corresponding properties of water. In comparing the properties of the refrigerants with those of water, two are of particular interest, viz., the normal boiling point and the latent heat. The normal boiling

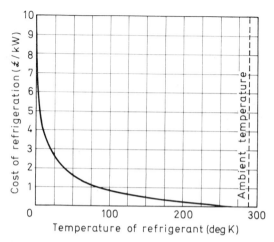

Fig. 2.1. Cost of refrigeration vs. absolute temperature.

points are very much lower than that of water, and the latent heats vary from 20% to 1% of the value for water. It is the low latent heat which is of particular interest as, particularly in the case of helium, this is one of the predominant reasons why low-boiling fluids require good insulation.

TABLE 1

SOME PHYSICAL PROPERTIES OF LOW-BOILING FLUIDS

	He^4	H_2	N_2	O_2	H_2O
Molecular weight	4	2	28	32	18
Normal boiling point, °K	4·2	20·4	77	90	373
Latent heat, kcal/kg at 1 atm pressure	6	106	48	51	540
Liquid density at NBP	0·125	0·07	0·8	1·15	1

In 1898 Dewar[1] demonstrated that an evacuated space between suitably silvered vessels forms an effective thermal barrier. Twelve years later Smoluchowski[2] showed that an effective insulator could be provided by combining fine powders with moderate vacua. Smoluchowski's work led to the widespread use of vacuum powder insulation which, for moderate low temperature applications, is unlikely to be replaced in the immediate future. An improved version of the vacuum powder insulant is a material consisting of a mixture containing equal quantities of insulating powder and fine metallic powder: the purpose of the metallic powder is to act as a reflector and reduce the passage of heat through the insulation. The major difficulty in using this material is in the prevention of stratification as its thermal efficiency is very dependent on the powder being mixed well.

The material known today as superinsulation was discovered first in 1951 by Peterson[3] at the University of Lund in Sweden. It did not arouse much interest at the time because of the widespread use of cheap powdered materials such as Santocel and Perlite. Superinsulation is a multilayer reflecting type of material with an exceptionally low apparent mean thermal conductivity. It consists of a number of highly reflective radiation shields interleaved with layers of insulating material and to be fully effective must be contained in a high vacuum environment. Only during the last decade has interest been shown in the use and development of superinsulation, and this is the direct result of the rapid advance in cryogenic engineering.

Cryogenic engineering is concerned with the production and use of very low temperatures, generally below $100°K$. It extends into a wide variety of engineering sciences including chemical, nuclear, electrical and aeronautical. The generation and use of very low temperature refrigeration necessitate the use of special techniques and special materials, one of which is superinsulation.

MECHANISM OF HEAT TRANSFER

The passage of heat from the ambient surroundings to a cold surface is caused by convective, conductive and radiant heat transfer. If the refrigeration maintaining the surface cold is to be preserved each of these mechanisms must be reduced or eliminated.

Convective heat transfer is caused by the temperature, hence density, of the air adjacent to the surface being different from that of the air remote from the surface. As a result, the air circulates, thus setting up currents. Because of this temperature difference, too, heat will pass from the surroundings to the cold surface by conduction through the air. Furthermore, the surface will receive heat by radiation from the surroundings, on account of the different temperatures again.

If the surface, which may be that of a container, is surrounded with an outer jacket and the air is withdrawn from the interspace, as shown in Fig. 2.2, these effects can be altered favourably. Normally, the thermal conductivity of a gas is independent of pressure, but when the mean free path of the molecules approaches or exceeds the width of the gas space, the thermal conductivity becomes proportional to the absolute pressure.[4] Thus, if a sufficiently high vacuum is maintained in the interspace the effect of conduction through the air can be ignored; in the case of air this requires a residual pressure of about 10^{-4} torr. The effective elimination of the air removes also the convective heat transfer.

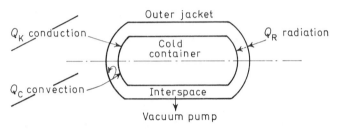

Fig. 2.2. Vacuum insulation.

Assuming the outer jacket to be at ambient temperature and the inner surface very cold, the transfer of heat between the two surfaces has been reduced to that by radiation only. Unfortunately, it is necessary to support the inner vessel from the outer one and this will, as a consequence, introduce heat inleak by conduction as shown in Fig. 2.3.

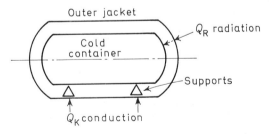

Fig. 2.3. Heat inleak.

With the conduction and convection effects more or less removed, the next step is to reduce the radiation effect.

The heat transferred by radiation can be expressed according to the Stefan–Boltzmann Law.

$$\frac{Q}{A_1} = \sigma \frac{1}{\frac{1}{e_1} + \frac{A_1}{A_2}\left(\frac{1}{e_2} - 1\right)} (T_2^4 - T_1^4) \qquad (1)$$

where Q = total heat flow watts
 σ = Stefan–Boltzmann constant
  $0\cdot533 \times 10^{-8}$ watts ft^{-2}deg K^{-1}
 A_1 = area of inner surface
 A_2 = area of outer surface ft^2
 e_1, e_2 = surface emissivities
 T_1, T_2 = surface temperatures °K
or

$$\frac{Q}{A_1} = E\sigma(T_2^4 - T_1^4) \qquad (2)$$

where E = effective emissivity.

Each of these equations relates to the heat transfer between two surfaces at different temperatures. If radiation shields are placed between the two surfaces the heat flux is reduced to

$$\frac{Q}{A_1} = \frac{E\sigma}{n+1}(T_2^4 - T_1^4) \qquad (3)$$

where n is the number of radiation shields.

If the radiant heat transfer is to be reduced to an acceptable level it is necessary to introduce a large number of shields. For the best performance the shields must be highly reflective, a property which is found in certain metals possessing extremely small surface roughnesses. Thus, thin metallic films can be used for this purpose. The temperature of each layer will differ from that of adjacent layers as expressed below and shown graphically in Fig. 2.4.

$$Ti = \left(T_2^4 - \frac{i}{n+1}[T_2^4 - T_1^4]\right)^{\frac{1}{4}} \qquad (4)$$

where Ti = temperature of ith shield °K
 n = number of layers
 T_1, T_2 = boundary temperatures °K.

If a number of layers or sheets of metal are placed side by side to form an effective radiant barrier, they will touch one another and act as a solid conductor thus defeating their object. This can be avoided partially by separating the layers with sheets of suitable insulating material of low thermal conductivity. Hence, by using a large number of radiation shields the radiant heat transfer is reduced but at the expense of additional conduction through the metal sheets and separating elements.

$$Q = \frac{kA(T_2 - T_1)}{L} \tag{5}$$

where Q = heat transferred   watts
       k = thermal conductivity   watts ft^{-1} deg C^{-1}
       A = contact area   ft^2
       T_1, T_2 = surface temperatures   °C
       L = thickness of insulator   ft.

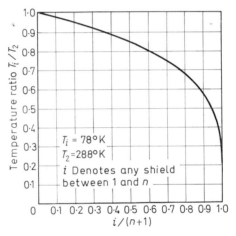

Fig. 2.4. Temperature distribution of floating radiation shields.

For ease of handling and application it is necessary that a thin insulator be used as a result of which, if the heat transfer is to be small, the contact area must be small also.

The above analysis results in an insulating system consisting of three principal factors: a high vacuum to eliminate heat transfer through the gas otherwise surrounding a container; a series of radiation shields to reduce radiation to a minimum acceptable level; spacers between the radiation shields to minimise solid conduction. Materials possessing these properties are known as "Superinsulants".

It should be realised that this simple analysis based on the Stefan–Boltzmann law does not strictly apply to a system consisting of multiple radiation shields separated by insulators; the layers of insulating material represent an absorbing medium, hence the heat transfer mechanism becomes complex and the mathematical solution even more complex.

TYPES OF SUPERINSULATION

Two principal types of superinsulation are used, viz., foil and paper material, and metal-coated plastics.

The foil or metal coating must be ductile for handling and shaping and be highly reflective. It must also be very thin because of the extreme anisotropy of such a system and the fact that in practice interlayer contacts are likely to occur. As the temperature of each metallic layer differs, the material from which the thin insulating sheets are made must possess a very low thermal conductivity. Moreover, the area of contact between the metal and insulator must be as small as possible.

The reflectors are made generally from aluminium and the insulators from woven fibrous materials such as glass fibre; aluminium possesses the desired combination of ductility and good reflective properties and the fibrous materials are good insulators. In addition metal-coated plastics are used such as aluminised Mylar. Mylar is a material made from polyethylene terephthalate. It is both strong and flexible, and is coated easily by vacuum deposition of metallic films.

It is of paramount importance that the area of contact between adjacent layers be kept to a minimum. In the case of the foil/fibre material advantage is taken of the weave as shown in Fig. 2.5. As the plastics material is not woven other means have to be found to give the same effect. This is achieved by crinkling the material as shown

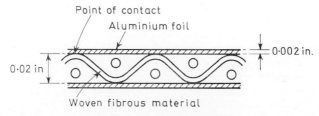

Fig. 2.5. Foil/fibre material.

in Fig. 2.6. The result in each case is that the contact area is small and the conduction path tortuous.

As a high vacuum constitutes an important part of a superinsulated system these materials have been chosen not only for their good thermal properties but also because their surfaces adsorb very little gas and they pump down easily.

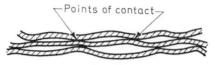

Fig. 2.6. Metal-coated plastics.

By way of comparison the density and apparent mean thermal conductivity of a few selected insulants are given in Table 2.

TABLE 2

SELECTED INSULANTS

Material	Density (lb/ft^3)	Thermal conductivity k ($Btu/h\ ft^3\ deg\ F/ft$)
Mineral wool	6	2×10^{-2}
Perlite	10	1×10^{-3}
*Aluminium foil/fibre glass	16	2×10^{-4}
†Aluminised Mylar	3	2.5×10^{-4}

*Aluminium foil	0·002 in thick
Fibre glass spacer	0·020 in thick
†Aluminised Mylar	0·00025 in thick

APPLICATION

When a superinsulant is tested in a calorimeter it is found to possess an extremely low heat flux. If this is to be preserved exacting methods of application must be used, otherwise the effect of using multiple reflectors and insulators may be destroyed.

The correct spacing of the layers must be achieved and optical windows and short circuits must be avoided.

If the packing density, which is determined by the spacing of the layers, is increased, it will result in increased areas of contact thus increasing the lateral conduction.

There are various ways in which superinsulation can be applied, and these, and the reasons for doing them, can best be understood by considering the insulation of a typical vessel such as the one shown in Fig. 2.7.

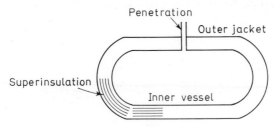

Fig. 2.7. Typical vessel.

Assuming that 40 layers are necessary the main vessel body can be covered in either of two ways depending on its size. Firstly, the material can be wound on until 40 layers are applied; it is important during this operation to ensure that the correct tension is maintained, otherwise the packing density will be wrong. Secondly, 40 sheets can be prepared to the correct dimensions and made-up in the form of a "blanket" which is then wrapped around the vessel shell. Insulation can be applied to the ends in a similar manner, that is, in the form of a blanket. This results in the main vessel body and ends being thermally isolated from the surroundings and the ends of the sheets connected together to complete the continuity of the insulant.

If the ends are butted together interlayer contacts will occur as will gaps between the two "blankets". Interlayer contacts will cause short circuits and gaps will act as optical windows so that the cold surface will see a region of high temperature. A thermally inefficient joint such as this will lead to a high rate of heat transfer and thus destroy the value of superinsulating.

So as to reduce the flow of heat through joints or discontinuities to a value compatible with that of the continuous material several methods are used. Figure 2.8 shows three typical ways in which this can be done. In Fig. 2.8(a), a simple overlap is used and this can be very efficient provided that the overlap is sufficiently long to minimise the conduction loss it introduces. Two difficulties are introduced in this method; warm and cold surfaces are in direct contact thereby introducing conduction along the layers, and a double thickness is formed which, if compressed to the thickness of a single "blanket", will increase the contact area and hence the lateral conduction. The second method, shown in Fig. 2.8(b), has advantages over the simple overlap method in that the layers in contact with one another are either warm or cold; this avoids heat leak by conduction along the layers. However, if the joint, which is now three times as thick as a single blanket, is compressed to a single thickness, increased lateral

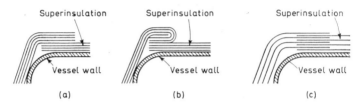

Fig. 2.8.(a). Simple overlap; (b) multiple overlap; (c) interleaving.

conduction will result. The third method, shown in Fig. 2.8(c), consists of interleaving individual layers and this offers the best performance. It does, however, introduce a double thickness which again may result in increased lateral conduction, if compressed. The outstanding penalty resulting from interleaved joints of this type is economic as it is an extremely time-consuming operation and can be justified only in the most exacting cases.

Discontinuities are treated in a rather different manner and insulated according to the temperature distribution in the system. Figure 2.9 shows a penetration and the method of applying superinsulation around it. In the first instance it is essential to know the temperature of each individual layer of the insulant; this can be determined approximately using eqn. (4). The temperature profile of the penetration or discontinuity is required also. With these data available it then remains only to match the individual layers to the points on the discontinuity having the same temperature.

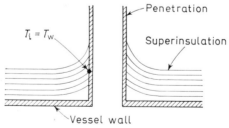

Fig. 2.9. Insulation around discontinuity. T_i is the temperature of an individual layer and T_w is the wall temperature.

During application the material must be kept clean and must not be damaged. If it is to be contained in a sealed high-vacuum environment some form of adsorbent must be introduced with it so as to avoid any pressure rise resulting from outgassing or diffusion of air through the container walls. After the insulated vessel has been assembled inside the outer casing it should be baked out at a moderate temperature and high vacuum to drive out trapped and adsorbed gases.

USES OF SUPERINSULATION

The present uses to which superinsulation is put are confined to the discipline of cryogenic engineering. This applies to any system where a liquefied gas, generally below $100°K$, is generated, stored or transferred.

Superinsulation is now making possible the long-term storage of liquid hydrogen and liquid helium. Vessels used for containing these fluids can now be made with a boil-off rate of less than 1% of the contents in 24 hours and this is achieved without the need to surround the vessel with a liquid-nitrogen-cooled radiation shield.

Transfer lines carrying liquefied helium, hydrogen, nitrogen and oxygen over long distances are now being insulated with superinsulation. This is a distinct asset in the case of pipelines carrying small liquid flows as the boil-off in transit could be extremely high. As stated previously even a low heat flux can result in a high boil-off rate because the latent heats of these fluids are so small. The ultra-low heat flow encountered with superinsulated transfer lines is therefore a useful tool in such cases.

New designs of very low temperature refrigerators are being developed to operate at temperatures from 2 degrees Celsius down to almost absolute zero. In pursuit of such temperatures new designs and components are demanded. A point of particular importance is that the efficiency of such a unit must be very high and when the factors contributing to efficiency are broken down into their component parts one of these is seen to be heat inleak which can be very significant. Superinsulation is being incorporated into these plants so as to reduce the loss of refrigeration at the lowest temperatures to an absolute minimum thereby increasing the efficiency of the cycles used.

In the production of rockets and manned space craft it is of vital importance to insulate certain parts. Insulation is used for two quite distinct reasons in space vehicles. Firstly, the vehicle will probably be powered with fuels such as hydrogen and oxygen which, for maximum utilisation of space, are stored in the liquid form. Secondly, the cabin of the space vehicle must be insulated to protect the occupants.

Once a space vehicle is outside the earth's atmosphere it will be contained in a high vacuum environment. Thus, the heat transfer resulting from convection and conduction through the air is eliminated leaving the mechanism of radiation only. If the vehicle can see the sun it will receive radiation but if it only sees space it will radiate to space in the same way that the earth's surface loses heat by radiation when ground frost occurs.

In order to prevent heat from entering the fuel storage containers, it is essential for the insulation used to have high reflecting properties; this has led to the universal use of superinsulation which is an exceptionally good reflector. It is important also to use a material of low bulk density which again leads to superinsulants. For similar reasons the cabin of the vehicle is insulated with superinsulation.

When a space vehicle takes off it is subjected to very high acceleration forces and it is very important that the insulant be able to withstand these. The material must not tear as this would cause optical windows and discontinuities. Moreover, it must be resilient enough to revert to its original shape once the force has been removed. Any retention in the compressed state would lead to increased contact area and undesirably high heat leak and loss of fuel. In the case of the cabin this may lead to a loss of internal energy.

As the weight of the equipment on a space vehicle is a serious penalty a material possessing a low thermal conductivity–density product ($k\rho$) is desirable and this is another favourable feature of superinsulation.

Apart from the more widespread use of superinsulation in refrigerators, storage systems, transfer systems and airborne applications, new applications are emerging in particular in the electrical engineering field. Research and development programmes are under way in the use of superconducting machinery and cryogenically cooled equipment and transmission lines.

The present method of transmitting power is by using a number of overhead cables which are generally large. Each of these cables passes a certain current at a high voltage usually of about 132,000 V. The limiting factor preventing a greater current being passed through the cable is its resistance which, measured over several miles, will be quite high. If more power is to be transmitted either more cables must be used or a more sophisticated system developed.

Certain materials possess the property known as superconductivity when cooled below the transition temperature peculiar to each material. Some of these are listed in Table 3.

TABLE 3

SUPERCONDUCTORS

Element compound	Transition temp (°K)	Critical field (Oersted)
Nb_3Sn	18	–
Pb	7·22	800
Hg	4·16	400
Al	1·175	106
Zn	0·93	45

In the superconducting state these materials have no measurable resistance but when a certain critical magnetic field is reached their resistance is restored. Nevertheless if the temperature is maintained below the transition point and the magnetic field below the critical, a very small conductor will pass a very high current. The maintenance of these temperatures requires the use of liquid helium as a refrigerant surrounding the conductor. In order to limit the loss of refrigeration to the surroundings, superinsulation must be used; any other form of known insulation would be completely uneconomical.

FUTURE TRENDS

It is difficult to predict how superinsulation will develop but work is in progress on the use of thinner materials as well as different types of materials. It is unlikely that aluminium will be replaced because it possesses excellent reflecting properties. However, other types of insulator are being tested in particular organic fibres, with results which appear encouraging. All the present forms of superinsulation are extremely effective and in the light of present technology there does not appear to be a demand for more efficient materials. This is not necessarily true of the application and it is in this field where new methods and techniques will emerge in the immediate future. An improvement in application, particularly at discontinuities or joints, could have a marked effect on the practical efficiency of a super-insulated system.

EFFECT OF SUPERINSULATION ON ENGINEERING TECHNIQUES

The use of superinsulation in cryogenic plants has led to a drastic reduction in heat inleak with the result that it has shown more clearly the losses created by components, such as valves. Valves are now designed to a high degree of thermal efficiency with a total loss of refrigeration at 10°K in the region of 1 or 2 watts. Pipelines terminate at warm boundaries with long thermal shunts and the minimum of material is used to support vessels and components.

REFERENCES

1. *Collected papers of Sir James Dewar,* Cambridge University Press, **1** and **2**, 1927.
2. M. Smoluchowski, *Bull. Int. de L'Academie des Sciences de Cracovie, Serie A* (1910) 129.
3. P. Peterson, *The Heat Tight Vessel,* University of Lund, Sweden (1951), Office of Naval Intelligence Translation, 1953, p. 1147.
4. William H. McAdams, *Heat Transmission,* 3rd Ed., McGraw Hill, 1954, p. 29.

3

A Mechanically Strong Thermal Insulator for Cryo-Systems

S. D. PROBERT and T. R. THOMAS
School of Engineering, University College, Swansea (Great Britain)
and
D. WARMAN
Department of Applied Physics, Welsh College of Advanced Technology, Cardiff (Great Britain)

INTRODUCTION

Multilayer superinsulations of aluminium foil have been developed for insulating cryogenic systems, mainly from thermal radiation.[1] However, they suffer from the disadvantage of low compressive strengths; mechanical loads can irreversibly reduce their insulating efficiencies by several orders of magnitude. With liquid refrigerant containers being subjected to large dynamic forces (as in spacecraft) it has become necessary (if solid supports and hence high conductive heat leaks cannot be tolerated) to supplement the superinsulation by load-bearing structures which have very high thermal resistances. The phenomenon of contact resistance provides a possible solution of this problem.

DESIGN PRINCIPLES OF MULTILAYER STACKS

When two macroscopically "flat" surfaces are pressed together, contact occurs only over very limited regions, because of the microscopic irregularities. If the surfaces are curved the true contact area becomes an even smaller fraction of the apparent contact area. Under high vacua ($\sim 10^{-6}$ torr) the passage of phonons and free electrons from one surface to the other under the action of a temperature difference is inhibited because the thermal carriers are restricted to flow through the asperity bridges. Thus a multilayer stack provides high thermal resistance. Thomas and Probert[2] made the following recommendations for increasing the thermal resistances per unit length of stacks:

(a) The thermal conductivity of the bulk material should be low.
(b) High surface hardnesses are required.
(c) The individual layers should be corrugated and stiff.
(d) A large number of thin layers is needed.

Of the materials tested the most suitable for making mechanically strong (in compression) thermally insulating stacks is the $2 \cdot 2 \times 10^{-3}$ in-thick Firth–Brown Staybrite F.S.T. stainless steel. A high value of yield strength, Y, to effective conductivity, k, is required, and is provided by the Staybrite stack, as shown in the Table. When the temperature is reduced, k decreases and Y increases, so that the figure of merit (Y/k) increases.

	Conductivity* k, under 140 psi load (10^{-2} Btu/ft^{-1} h^{-1} deg. F^{-1})	Yield strength* Y (10^3 psi)	Figure of merit $\dfrac{Y}{k}$ (10^4 psi ft h^{-1} deg F Btu^{-1})
Staybrite stack, $2 \cdot 2 \times 10^{-3}$ in layers	22	59	27
Tufnol "Carp" phenolic resin, $7 \cdot 0 \times 10^{-3}$ in layers	3·8	5·0	13
Solid Mylar	8·8	10	11
Brass stack, $2 \cdot 3 \times 10^{-3}$ in layers	48	23	5·0

* Measurements made at 59°F.

STABILITY OF MULTILAYER STACKS

Stacks of flat discs have high compressive strengths; once the initial layer undulations are pressed out the effective Young's moduli are within two orders of magnitude of those of the bulk material, and increase with load.[3] Moreover, there is no likelihood of the collapse of short stacks of flat layers under loads applied normally to the disc faces (cf. Euler's theorem). Longer stacks (with length-to-section ratios greater than 10), however, especially when subjected to lateral thrusts, may fail as a result of disc slip even under high axial loads. Thus layers of different shapes were considered. Cost precluded adopting thin annular rings of corrugated stainless steel (as are often used for pressure capsules). The punching of cones from the thin sheet resulted in jagged inner peripheries, so that when stacked the cones tended to become embedded in adjacent layers, and this led to decreased thermal contact resistances. Thus the *caltrop* elements (shown in Fig. 3.1(a)) were devised and punched out of the 2·2 × 10^{-3} in-thick Staybrite sheet. Low-thermal-conductivity supports and stacks of this type are the subject of a UKAEA patent application.[4]

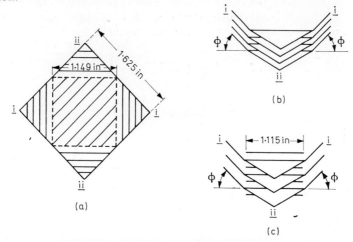

Fig. 3.1. Elements of laterally strong stacks. (a) Caltrop, – – –, line of bend. i, bent up, ii, bent down; (b) stack of caltrops; (c) stack of alternate flat and caltrop layers.

However, long stacks of caltrops, as shown in Fig. 3.2, distorted under axial loads and eventually collapsed (see Fig. 3.3); this is attributed to the line of bends not being infinitesimally sharp. So flat squares of side 1·115 in of the same material were sandwiched alternately between the flat square areas of caltrops (as shown in Fig. 3.1(c)); the curve of compression against applied load (Fig. 3.3) for such a stack shows that no slip was experienced. The compressions (chosen as zero at 50 lbf) for similar loads were less than an order of magnitude greater than those for a stack of flat discs of equal area and of the same material. At the higher loads (above 7×10^2 lbf) the compression becomes a linear function of the logarithm of the applied load, as predicted by theory.[3]

For improved lateral stability the protruding jaws of the end blocks for the stacks were shaped to accommodate changes of wing angle (ϕ) that occur when the stack is squashed by loads applied only on the flat square areas.

Fig. 3.2. Stack of caltrops.

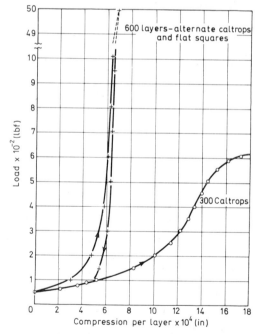

Fig. 3.3. Axial load–compression curves for two stacks.

BASIC DESIGN OF THE THERMAL INSULATOR

The mechanically strong thermal insulator consists of six stacks as described above (of alternate flat square and caltrop layers between shaped end-pieces). The prototype is shown in Fig. 3.4, and schematically in Fig. 3.5. The three upper stacks are each of 500 layers and the three lower stacks of 650 layers. The six stacks are arranged so that in the loaded prototype they form three sets of collinear pairs, with axes at 30° to the vertical. Any movement of the central

Fig. 3.4.

34   THERMAL INSULATION

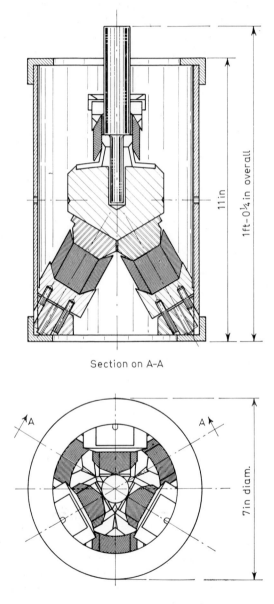

Fig. 3.5. Schematic diagram of the prototype insulator.

boss relative to the outer casing automatically results in a restoring force system in the reverse direction. As the axial load is increased, the conductance of the three lower stacks increases, whereas the load on the upper three stacks decreases, leading to a reduction in their conductance. Figure 3.6 shows that the total effect is a relatively small change in the solid conductance of the support over a wide range of applied loads, for temperature differences of practical interest.

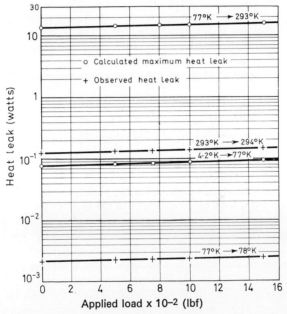

Fig. 3.6. Solid conductance heat leakage through the prototype support (six stacks) as a function of the applied load, for various temperature differences.

The designed maximum axial static load on the support was $1 \cdot 25 \times 10^3$ lbf, though this may be far exceeded as a result of mechanical shock. For use under average static load conditions the three upper stacks were preloaded to 5×10^2 lbf and the three lower ones to $8 \cdot 5 \times 10^2$ lbf. Initially it was hoped to use far lower preloadings, and hence to obtain much higher thermal resistances than those indicated by Fig. 3.6. However, for axial loads cycled up to 5×10^4 lbf and transverse loads up to $1 \cdot 25 \times 10^3$ lbf applied to the boss, the outside shell being fixed, the prescribed preloadings had to be used to eliminate any likelihood of buckling couples on the stacks resulting in their axes becoming S-shaped instead of straight, and eventually in collapse. If the system has only to withstand lower forces, the preloadings can be reduced and lower thermal conductances through the support will result.

Each stack was prestressed by two high-tensile bolts before being introduced into the shell assembly. These prestressing bolts were removed from the split lugs (not shown in Fig. 3.5) before withdrawal of the screwed fixing pins positioning the inner boss; the locating screwed holes in the outer casing and depressions in the boss for the locating pins are shown schematically in Fig. 3.5.

Three holes (of diameter 3·0 in) were cut in the outer casing of the stainless-steel tube (which is 0·201 in thick). These both lighten the structure (the total weight of the prototype being 39·6 lb) and facilitate the installation of the stacks and removal of the prestressing bolts. A further three lightening holes could be drilled into the central boss, so reducing the weight of the assembly by a further 3 lb approximately, without an appreciable diminution of strength. All components of the support would normally be of stainless steel because of its low thermal conductivity, with polished surfaces to reduce radiation exchange.

The data presented in Fig. 3.6 for the variation of thermal conductance with extra load applied axially to the shaft of the prototype are for the "run-in" state; the layers in each stack had become bedded down as a result of load cycling, and the thermal resistance had decayed considerably from the initial value on assembly. The way in which dynamic loads lead to a decrease in the resistance (both electric and thermal) of stacks has been examined elsewhere.[3]

The data (Fig. 3.6) for conductive heat leaks through the support under various applied loads are higher than will be encountered in practice between liquid-helium, liquid-nitrogen and room temperatures for two reasons: (a) mean "contact" conductivities for the whole of the considered temperature ranges are used, and (b) the graphs of thermal conductance per degree temperature difference across a contact under a specified load against temperature are concave upwards for stainless steels (deduced from refs. 6–8). This is an interesting trend in view of the graphs of thermal conductivity against temperature for stainless steels being concave downwards[9] over the same temperature range, below 500°R, *i.e.*, there is a difference between the temperature behaviour of the bulk and the surface phenomena.

Although the basic principles of the design of the insulator are fixed, experiments are being made to elucidate the optimum roughness for materials of a given (but very high) hardness. These experiments will permit further reductions in the heat leaks through the support. For the same reason the use of thermoelectric heat pumps[10] and the thermal rectifying action[11,12] of certain pressed contacts are also being investigated. It is desirable to be able to predict the thermal resistance of a pressed contact from a knowledge of the surface topography and hardness; this is being attempted using a Talysurf instrument fitted with a digital analyser feeding data to an IBM 1620 computer. The temperature variation of thermal contact resistance, however, is as yet far from predictable, but progress is being made by investigating the dependence of the surface hardness of materials on temperature.

ACKNOWLEDGEMENTS

Financial support for the work described was derived from the UKAEA, under extra-mural contract, and from the Science Research Council. Thanks are due to Stewarts and Lloyds Ltd. (Cardiff) for various steel tubes, and to Tufnol Ltd. for providing specimens.

REFERENCES

1. R. H. Kropschot, *Cryogenics*, **1** (1961) 1.
2. T. R. Thomas and S. D. Probert, *Int. J. Heat Mass Transfer*, **9** (1966) 739.
3. S. D. Probert and M. C. Jones, *J. Strain Analysis*, **1** (1966) 283.
4. S. D. Probert (assigned to UKAEA), Brit. Pat. Appl. **53336/65**.
5. M. C. Jones, *M.Sc. Thesis*, University of Wales, 1965.
6. R. Berman, *J. Appl. Phys.*, **27** (1956) 318.
7. R. P. Mikesell and R. B. Scott, *J. Res. Nat. Bur. Standards*, **57** (1956) 371.
8. T. R. Thomas and S. D. Probert, *TRG Report* 1013 (*R/X*), UKAEA, 1965.
9. J. de Nobel, *Physica*, **17** (1951) 551.
10. C. B. Thomas and S. D. Probert, *Brit. J. Appl. Phys.*, **15** (1964) 1120.
11. G. F. C. Rogers, *Int. J. Heat Mass Transfer*, **2** (1961) 150.
12. J. S. Moon and R. N. Keeler, *ibid.*, **5** (1962) 967.

4

Low-temperature Applications of Expanded Polystyrene and Expanded Ebonite

H. S. F. BAKER
Formerly Onazote Insulation Company, London, S.E.1 (Great Britain)

INSULATION

The applications of expanded polystyrene and expanded ebonite to be described in this chapter are, of course, as low-temperature insulation materials, and it is therefore fitting as well as convenient to commence with a definition of the term Insulation.

Insulation has been defined* as the method employed to restrict the operation of a law of nature whereby heat flows from a warm body or space to a colder one. In carrying out this function the insulation is exposed to the effects of a second law of nature, which is that water vapour will endeavour to flow from a region of higher vapour pressure to a region of lower vapour pressure.

It follows then that the most important characteristics of insulation materials, from the purely scientific point of view at any rate, are (i) low thermal conductivity (k) and (ii) minimum permeability to water vapour.

In general, materials consisting of a multiplicity of either fibres or minute air cells present the greatest resistance to heat transfer by convection. The earliest known method of thermal insulation employed was the utilisation of animal skins and furs for clothing and for tents; and there is evidence that the Phoenicians, Greeks and Romans had discovered that the outer covering of a variety of oak tree (Quercus Suber) growing in profusion on Mediterranean shores, had features peculiar to itself which prompted these people to use the bark for the retention of heat and the exclusion of moisture.

Towards the end of the nineteenth century two events took place which simultaneously brought about the large-scale use of natural cork. First of all it was found that, by breaking the cork bark into small granules and then subjecting it to steam heat under pressure,

---
* British Standard Code of Practice No. 406, 1952.

it was possible to obtain a block which had very high mechanical strength coupled with good thermal efficiency and a reasonable degree of resistance to water vapour ingress. By comparison with charcoal, pumice, seaweed and sawdust, corkboard was of extremely high quality. Almost coincident with this development was the appearance and growth of mechanical refrigeration.

In view of the high cost of extracting heat, the efficiency of thermal insulation is of the utmost importance. Corkboard remained for many years the outstanding low-temperature insulation medium, but inevitably a constant search went on during this period for less expensive materials with, if possible, even better properties.

Among the new products which have emerged are expanded polystyrene and expanded ebonite. These are cellular materials in which the cells do not intercommunicate, in contradistinction to the so-called "sponge" materials in which there is communication between the cells. Other plastics have also been successfully "expanded", including synthetic rubber, phenol/formaldehyde resin, urea/formaldehyde resin, cellulose acetate and polyvinyl chloride, but the cost of the basic raw materials or inherent defects in their physical characteristics make some of them of little interest to the refrigeration industry.

The desirable characteristics of insulation materials are:

1. Low thermal conductivity.
2. Resistance to water absorption.
3. Resistance to combustion.
4. Sterility in order to prevent the growth of living organic spores and fungi.
5. Ease of cutting and shaping.
6. Resistance to settlement (this is obviously important where vertical surfaces are insulated). If a material which is subjected to vibration tends to pack down, a void will ultimately appear at the top, permitting a rise in heat transfer.
7. Low weight per cubic foot.
8. Easy availability.
9. Resistance to vermin.
10. Low cost.

No pretence is made that this list is in order of importance. Indeed, from the purely commercial aspect, it might be reckoned that low cost would be very near the top, but of course the cheapest material is not always the best. Different characteristics take on different degrees of significance as between one application and another. If, for instance, an insulant is to be used to support pipework in place of brackets its structural strength must be taken into account, while for the insulation of a refrigerated vessel, which may never be subjected to mechanical weight or damage, this aspect is of less importance.

APPLICATIONS

The food industry is perhaps the largest consumer of expanded polystyrene and expanded ebonite low-temperature insulation. As the general standard of living has risen so has the demand for better quality and greater variety in all matters of diet. Not many years ago it was common to see in this country vast quantities of fresh fruit either destroyed or given away during the high season of maturity because adequate means of long-term storage were scarce and expensive. This not only meant a loss of money to the producer, but also limited to a few short weeks the period during which the consumer could enjoy such produce. It is commonplace now to be able to eat dessert apples, strawberries, garden peas and beans throughout the year, simply because they have been stored and preserved in perfect condition over long periods by refrigeration.

Cold-stores for meat, poultry, cheese, butter, fruit, etc., all present their individual problems, depending on the temperature and humidity conditions of storage necessary. They do, however, possess one common factor, which is that the insulation should be as efficient as possible in order to keep running costs to the lowest level.

An excellent example of the combined use of expanded polystyrene and expanded ebonite is a cold-room built, by Northern Sea Foods, to operate at $-10°F$. The floor is insulated with 6 in of expanded ebonite to give good water vapour and heat resistance and to carry the heavy loading of frozen produce. To the walls were applied 2 in of expanded ebonite to provide a vapour barrier coupled with high heat transfer resistance, followed by 5 in of expanded polystyrene, the latter being supercoated with cement as mechanical protection. The ceiling is insulated with 7 in of expanded polystyrene, with a vapour barrier applied over the hot face.

It is becoming increasingly common practice to erect expanded polystyrene as the ceiling insulation to large cold-stores and to omit any form of internal finish. The insulant is white and, therefore, suitable as a decorative finish.

Many of the largest apple stores in Britain are insulated in a similar manner to the above, though the insulation thickness is reduced because of the higher operating temperature and, in this case, the insulation is lined internally with suitable gas-sealing material.

From the storage area to the retail shop it is necessary that the produce be maintained in good condition and this has brought about a high usage of expanded polystyrene for the insulation of road transport vehicles. Its extremely low weight, coupled with excellent thermal properties, has made expanded polystyrene the automatic choice of most transport operators. Pay loads are much increased in respect of both weight and cubic capacity, and the insulant shows no sign of deterioration over very long periods.

Because of its efficiency and low cost, expanded polystyrene is also the most commonly used insulant for refrigerated counters and display cabinets in shops.

An interesting ultra-modern abattoir is at present under construction in Cardiff. Very large areas of chilled space are being provided for the housing of the carcasses, and private chill rooms for the use of the wholesale distributors. Deep freeze conditions are to be available for the storage of the offal. The floors of all refrigerated areas are to be insulated with expanded ebonite while walls and ceilings will be in expanded polystyrene. Expanded ebonite has been chosen as the insulant for refrigeration pipework and vessels throughout.

A very large number of fish trawlers in current service have their fish holds insulated entirely with expanded ebonite. Even where other insulants have been used for bulkhead insulation, it has been found expedient to employ expanded ebonite for the deckhead and the tank tops because of its ease of application and load-bearing quality.

It can be seen from the above examples how great a contribution the two expanded materials under discussion make to the food industry from farm or sea to consumer.

Most of what has been described so far is, however, fairly commonplace as compared with the more dramatic examples of the uses of these materials to be found in the chemical and petroleum industries.

Some years ago enormous reserves of methane gas were discovered in Algeria, far in excess of any possible local demand. The only conceivable method of exporting this gas to areas of high demand was to reduce its volume by liquefaction so that it could be transported by sea, and a plant was laid down at Arzew, on the Algerian coast, for this purpose. Trial shipments were made to Canvey Island, in Essex, where the liquid was off-loaded and processed to produce methane gas for blending and distribution through the National Grid. The temperature of the liquid has to be held at or below $-258°F$ and, as can be imagined, the requirement of a super-efficient insulant soon became apparent. Expanded ebonite was chosen for the pilot scheme and proved to be so successful that it was later specified for the full-scale commercial plant which now contributes largely to our gas supplies.

Several factors combined to cause expanded ebonite to be chosen. In the first place it is possible to machine it so accurately to half-round sections that a perfect fit to a pipe is obtained. In addition it can stand up to the extremely low surface temperature of the pipe without deforming, except in respect of the natural thermal contraction which affects all materials.

The insulation was applied in multi-layers up to a total thickness of six inches, and a gradual contraction made layer by layer as the interface temperatures approached the ambient. For such an exacting application, of course, special methods had to be employed, so that far more elaborate outer coverings were applied than would be the case on pipework operating at higher temperatures.

Despite the fact that expanded ebonite is virtually totally resistant to water vapour ingress, it was considered expedient to enclose the insulation in grease-impregnated tape prior to the application of the final protective finish, which comprised either bitumen roofing felt and chicken wire, or aluminium sheeting, depending on the aesthetic requirements of the site. The same system has subsequently been adapted to liquid methane lines at Arzew, with similar success.

Propane is another relatively cheap form of fuel and, for reasons similar to those described above, is only transportable economically in liquid form.

Expanded ebonite was used in large quantities to insulate six on-deck tanks on the vessel "MUNDO GAS", built in Norway. This application was interesting because of the conditions to which the insulation would be subjected on long ocean voyages. Various finishing coats were considered and rejected because of their limited resistance to saline atmosphere, and it was finally decided to coat the insulation externally with polyester glass fibre laminate. This operation was successfully carried out and other vessels have since been treated in a similar fashion.

A butadiene extraction plant was built several years ago at the Esso refinery, Fawley, at a cost of £10,000,000. Many large solution exchangers and settler cylinders were insulated with expanded ebonite.

EXPANDED POLYSTYRENE AND EBONITE

The same material has been used on large Horton spheres for the storage of liquid ammonia, including one which is possibly the largest vessel in the world used for this purpose. It is a 60-ft-diameter sphere at the premises of Fisons Fertiliser Division, Stanford-le-Hope (Fig. 4.1).

Fig. 4.1. Horton sphere, 60 ft in diameter.

Expanded insulation materials also play their part in the field of recreation. The majority of ice-skating rinks built in this country since the war have this form of insulation as the barrier between the sub-floor and the refrigeration coils. The coils are coupled to refrigerant mains which run along one side of the rink in a channel called a header trench.

At a typical modern ice-rink expanded ebonite is used to line the walls of the trench to provide good heat and vapour resistance. Expanded polystyrene is then used in board form to fill between the pipes and, finally, expanded polyurethane is foamed *in situ* to fill the remaining spaces. The final result is a thoroughly efficient insulation which would be impossible to achieve by any other means.

5

Applications of Rigid Polyurethane Foam

W. J. WILSON

Dyestuffs Division, Imperial Chemical Industries Ltd., Manchester
(Great Britain)

The use of cellular materials as insulants is of long standing and many of the traditional insulating materials owe their properties to the fact that they have structures which enclose air and prevent it from escaping readily under the influence of thermal convection.

It is not surprising, therefore, that the ability to produce cellular products from synthetic materials has resulted in such products being used in insulating applications for which the traditional fibrous or naturally occurring cellular insulants had hitherto been used. Such synthetic products include cellular PVC, expanded ebonite, expanded polystyrene and polyurethane foam, and it is with the applications of the last-named that this chapter is concerned.

The early types of polyurethane foam were produced with carbon dioxide in the cells, many of which were interconnected, thus allowing the carbon dioxide to be replaced by air almost immediately after manufacture. The carbon dioxide also diffused quite rapidly through the walls of those cells which were completely closed and was soon replaced by air which diffused inwards. Such foams, while functioning as excellent insulation materials, were limited in efficiency by the fact that the essential insulant was air and that a portion of this at least could move about fairly freely within the foam.

Modern polyurethane foams can be produced having cells nearly all of which are completely non-interconnecting but, in addition to this advantage, the cells at the time of manufacture are filled with a chlorofluorohydrocarbon such as trichlorofluoromethane, which may or may not be accompanied by carbon dioxide. If the surface of the foam is exposed to the atmosphere, air will steadily diffuse inwards through the cell wall until its partial pressure is equal to that of the atmosphere. The chlorofluorohydrocarbon, however, is unable to diffuse through the cell wall and remains permanently trapped there.

Since it has a thermal conductivity much lower than that of air, it forms an extremely effective barrier to the transmission of heat across the cell, so that the thermal insulating properties of such foams are greatly superior to those of their predecessors.

The course of the diffusion of air inwards may be followed by studying the thermal conductivity of the foam, which slowly rises from an exceedingly low value to a somewhat higher, but still low, equilibrium figure, which then appears to be maintained indefinitely (Fig. 5.1).

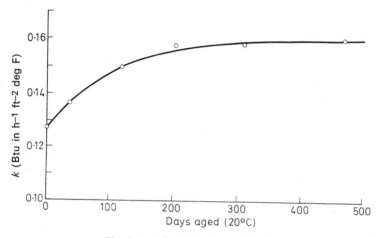

Fig. 5.1. Effect of ageing on k value.

In this connection, it may be noted that if the surface of the foam is completely sealed so that air cannot enter, the initial thermal conductivity will be maintained indefinitely; even partial sealing will delay the attainment of equilibrium very considerably.

The use of a chlorofluorohydrocarbon in modern polyurethane foams has not only reduced the thermal conductivity of these below that of the older types but has resulted in an insulant having thermal properties superior to those of any of the traditional insulating materials (Table 1).

There are a number of ways in which rigid polyurethane foam may be produced, but they are all based on the reaction of a polyhydroxy compound or "resin" with a polyfunctional isocyanate, additional agents such as catalysts and chlorofluorohydrocarbon blowing agents being used to regulate the course of the reaction. The reaction between the resin and isocyanate results in the production of a polymer which becomes more and more viscous and finally solid as the reaction proceeds. A considerable amount of heat is generated in the process and this serves to vaporise the chlorofluorohydrocarbon; bubbles of the vapour are trapped by the rapidly solidifying polymer and a mass of rigid foam is the result.

TABLE 1
THERMAL CONDUCTIVITY OF INSULATING MATERIALS

	Density (lb/ft^3)	k ($Btu\ in/ft^2\ h\ deg\ F$)
Cork	8	0·28
Fibreboard	16–20	0·37–0·45
Glass fibre	1·5–3	0·25
Expanded polystyrene	1·2	0·23
Expanded PVC	2·5	0·23
Expanded ebonite	4	0·20
Phenol/formaldehyde foam	3·5	0·26
Rigid polyurethane foam	2·1	0·12–0·16

It is important that the relative amounts of resin and isocyanate should be accurately maintained by means of carefully adjusted metering pumps and that all the materials involved should be mixed together thoroughly. This may be achieved by high-speed mechanical mixing, by causing jets of each component to impinge on each other at high pressure or by inducing turbulent flow in a small mixing chamber at low pressures. The latter allows very compact equipment to be made for the convenient application of foam by hand.

Sometimes, a part of the resin is reacted with the isocyanate to produce a half-way stage in the process and this pre-polymer is then reacted with more resin to produce the final foam.

Rigid foam may be produced in a number of different forms, for example, it may be made in large slabs, by allowing the mixed resin and isocyanate to fall on to a moving conveyor, where it rises to produce a continuous slab which may be cut up and sliced to give rigid foam in sheet form or cut into various shapes for special purposes. A second method is to allow the resin and isocyanate to mix intimately in a portable gun from which a stream of unfoamed mixture may be ejected at will by operating the gun trigger (Fig. 5.2).

By this means, the chemicals may be injected through relatively small openings into closed spaces such as refrigerator cabinets or building panels, and allowed to expand to about 30 times their original volume and so fill the spaces completely.

By modifying the gun slightly, the liquid may be sprayed on to a surface where it will react and expand to produce a highly adherent insulating layer.

By the use of a combination of the spraying technique with a moving conveyor and ancillary equipment, it is possible to produce continuous sheets of rigid foam of any desired thickness having facing materials already attached.

A further method employs, in addition to the materials already mentioned, a second chlorofluorohydrocarbon having a much lower boiling point than that of the one normally used, so that the foam emerges from the mixer in the form of a froth, which then continues to expand by the normal foaming reaction.

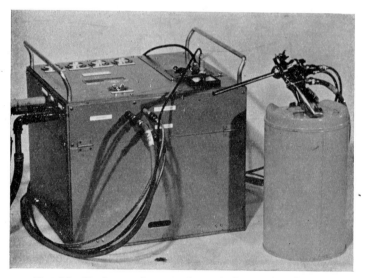

Fig. 5.2. Equipment for preparing resin–isocyanate mixtures.

All these methods are applicable to the use of rigid foam for insulation or constructional purposes and the choice of any particular method will depend on its convenience and suitability for the purpose and on the economics of the situation. Polyurethane foam not only possesses a very low thermal conductivity, thus enabling a thinner layer of insulant to be used, but its ability to be foamed in place, where appropriate, means that jointless insulation may be achieved with an appreciable reduction in heat losses. In addition, it adheres strongly to many common constructional materials, so that in many cases it acts as its own adhesive. By virtue of this bonding property, coupled with its high mechanical strength, the foam confers a marked degree of stiffness on structures in which it is incorporated, so that it is often possible to reduce the thickness of the facing materials to which it is bonded and thus achieve appreciable economy.

Taking all these points into consideration—high thermal insulation, self-bonding and economies in materials—it is often found that the apparently high cost of polyurethane foam per cubic foot is more than offset by savings in materials and labour, so that the overall cost of a job is reduced and its use becomes an economical proposition. It must be emphasised, therefore, that it is the cost of the finished job which must be taken into account rather than the unit cost of the insulant.

Applications of polyurethane foam as an insulant are to be found in a wide variety of industries. In the refrigeration industry, for instance, both domestic and commercial refrigerator and deep freeze manufacturers have realised the advantage of being able to utilise a thickness of insulant less than that commonly employed with conventional materials, so enabling a larger internal volume to be achieved without increasing the external dimensions of the refrigerator. Added to this, a complete refrigerator may be fully insulated in 15 seconds, simply by injecting the liquid polyurethane mixture through two small holes in the outer casing, a process which can be made semi-automatic and which is eminently suited to production-line techniques (Fig. 5.3).

Just as the foam may be injected into a refrigerator cabinet, so, in refrigeration applications on a larger scale, it may similarly be moulded, in contact with a suitable facing material, into slabs several feet in length and width, and 3 in or so thick. The low density of the foam—about 2 lb/ft^3—enables such large blocks to be handled without difficulty, and completely insulated cold-stores are being built up in this way.

In the building industry in general, as distinct from the rather specialised building of cold-stores, polyurethane foams have been used for several years not only as a means of providing thermal insulation but as integral parts of structures, such as panels and actual house walls.

The facade of the Chevron Hilton Hotel in Sydney (Fig. 5.4), for example, embodies 30,000 ft^2 of curtain wall panelling, built up from an outer facing of enamelled steel sheet and an inner facing which forms the internal wall surface. The space between the two facings was filled in the factory with rigid polyurethane foam by holding the panel securely between the platens of a press and injecting liquid foam material through a small injection hole. The foam expands almost immediately, displacing air from the interior through small breathing holes drilled near the corners. The amount of foam injected is accurately measured automatically by the metering equipment so that

Fig. 5.3. Complete refrigerator insulation.

complete filling of the panel is ensured, very little foam being wasted by exudation from the breathing holes. These panels have been in place now for about 5 years.

In this way, building panels of almost any size or thickness can be made, ranging from those having 3–6 in of foam for the construction of cold-stores to more elaborate structures faced with mosaic tiles, thus combining a decorative effect with excellent thermal and structural properties. Such panels are widely used today in shop fronts, office buildings and similar locations and between these extremes we have a whole gamut of panels using metal, hardboard, laminated plastics, asbestos and many other materials as facings.

Fig. 5.4. Facade of Chevron Hilton Hotel, Sydney.

By stacking several panels together in one press, a rapid rate of production may be achieved, as in the case of panels required for a large factory in Canada (Fig. 5.5). The size of the panels may be judged from a view of the finished building (Fig. 5.6).

Although a press is convenient as a means of holding the empty panels flat during the filling operation, it is not essential, and any means of jigging the panel facings rigidly can be employed, so that one is not limited by the sizes of press which are commercially avail-

THERMAL INSULATION

Fig. 5.5. Building panel production.

Fig. 5.6. Building panels after erection.

able. An example of this technique is shown in a factory-built house in which each wall, other than party walls, is a single panel some 24 ft long and of storey height (Fig. 5.7).

These steel-framed panels, faced with appropriate materials for their inner and outer surfaces, are mounted between two large hinged leaves forming a jig which holds them firmly while foam is injected from a number of points. The foam finds its way into every corner of the frame, bonding the whole into an integral structure and sealing

Fig. 5.7. Application of storey-height building panels.

all the joints. Gaps are left at appropriate places for windows and doors which are fitted when the panel comes out of its jig and the result is a complete wall of storey height. Houses can thus be erected very rapidly and there are now a large number being put up all over the country.

Steel panels filled with rigid foam are used extensively in Italy for the construction of schools, restaurants and hospitals. The panels are of standard dimensions and different types can be fitted together rapidly to produce buildings of almost any size or shape provided that they fit the standard module.

Figure 5.8 shows a hospital in Venice which has been built in this way, and although the outside may look a little austere, the interior (Fig. 5.9) belies the comparatively simple method of construction.

Fig. 5.8. Hospital in Venice.

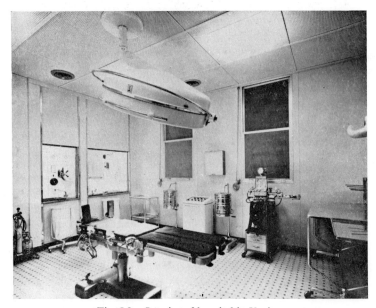

Fig. 5.9. Interior of hospital in Venice.

By the employment of similar techniques, the walls of a lightweight holiday chalet (Fig. 5.10) were made from sheet steel enclosing 6 cm of rigid polyurethane foam. These have been used in the Italian Alps for 5 years where they have been exposed to snow loads up to 4 metres deep. The thermal insulation is such that one 2-kW heater maintains the internal temperature at 25°C when the outside temperature is $-30°C$.

Fig. 5.10. Lightweight holiday chalet.

As indicated earlier, rigid foam may be applied by spraying; this method of application has been used mainly for the insulation of chemical plant and storage vessels. Thus, a large chemical tank (Fig. 5.11) has been protected from solar heat by the spraying of a "sunshade" of polyurethane foam over its upper surface, a neat and extremely efficient thermal insulation being provided by a rapid and convenient method of application.

This technique has also been applied to buildings. The roof of a church in Montreal (Fig. 5.12) has had a layer of polyurethane foam sprayed on to its roof to serve as a thermal insulant, the foam being subsequently covered with a layer of thin plastics roofing sheet.

It will no doubt have been observed that the examples of sprayed foam illustrated have involved large structures, where the efficiency of the process, in terms of amount of foam actually applied to the structure compared with that delivered by the gun, is high; in other words, the greater the target area the less wastage of material which is inevitably sprayed beyond its edges. It is therefore of interest to note that even relatively small items can be sprayed with foam economically, in spite of appreciable wastage by overspray, if the economics of a system as a whole are considered rather than just the cost of applying the foam in comparison with other materials.

It might, for example, be considered that polyurethane foam at a price of about 9 shillings per cubic foot would stand little commercial chance against conventional insulants at, say, 3 shillings, when used

Fig. 5.11. Application of polyurethane foam "Sunshade".

Fig. 5.12. Polyurethane foam applied to church roof.

for insulating ducting conveying heated air in domestic central heating systems. Yet at least one Gas Board in this country has adopted factory-sprayed polyurethane foam insulation for their ducted hot air system because the overall outlay is reduced by:

(a) The more rapid application of the insulation under factory conditions.
(b) The very considerable reduction of work on site.
(c) The saving in materials by their being able to use lighter gauge metal for ducting, since the rigid foam layer adds considerably to the strength of structure.

The individual sections of ducting in this case were not large and even when it was arranged for several sections to be sprayed at the same time the overspray losses were appreciable. Nevertheless, the financial savings achieved have been sufficiently great to justify completely this method of insulation, and this serves as an excellent illustration of what was said earlier: it is the cost of the finished job that matters and not that of the insulation alone.

The use of polyurethane foam in the form of sheets or other shapes which are cut from a large block of foam is more widespread in the United States than in this country, since the American manufacturers developed their rigid foam technology using the techniques already established for flexible foam production, *i.e.*, producing large slabs of foam by a continuous process on large machines built in factories where the toxic hazards of the tolylene diisocyanate used could be overcome by the provision of adequate ventilation. In Britain, and later in Germany, the development of the very much less toxic diisocyanatodiphenylmethane (MDI) has led to the more widespread use of the injection and spraying techniques, the scope of which is not limited by any need for special ventilation. Nevertheless, a considerable amount of rigid foam in block form is made in this country, and is used for general insulation applications, especially in the construction of cold-stores and in the manufacture of certain types of building panel where the injection technique would not be appropriate.

The techniques described so far have all involved two separate processes, viz. (a) producing the foam in liquid or expanded form and (b) applying suitable facing materials to it, or injecting it between prepared facings. A considerable amount of development work has therefore been done with the object of applying facing materials to foam by mechanical means while the latter is still reacting and thus reducing the complete process to one automated operation.[1] To date, this has been applied to flat objects such as building panels which are economical to produce in long production runs.

The principle of this method is shown in Fig. 5.13. It consists essentially of an endless conveyor moving over a fixed lower platen, above which is suspended a second conveyor moving over a platen which is free to rise or fall in a vertical direction. If the foam sheet is to be covered with a flexible facing such as paper or a decorative plastics film, these materials are fed from two rolls along the upper and lower conveyor and are transported through the machine in this way. At the

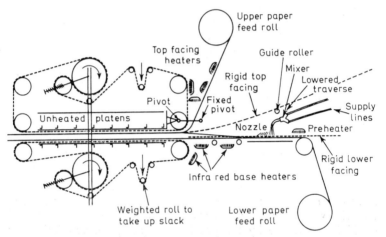

Fig. 5.13. Continuous production of building boards.

front end of the machine a foam spray gun is mounted so that it is made to traverse backwards and forwards across the lower facing as it moves into the machine; the foam is thus deposited on the lower facing and commences to expand. At the point where expansion is substantially complete but the surface of the foam is still tacky, the upper facing is brought into contact with the foam and the complete sandwich passed between the platens to emerge as a continuous length of composite board which may be cut automatically into suitable lengths.

It is important to note that in this process the function of the platens is not to provide a fixed gap in order to regulate the thickness of the resultant laminate; the upper platen floats on the foam and serves to ensure uniform contact between the upper facing and the foam core. In this way the direction and rise of the foam is maintained perpendicular to the plane of the board and is not distorted by being forced through a fixed gap; as a result, the mechanical properties of the final laminate are considerably improved. The thickness of the board is determined by the amount of foam being laid down which is in turn regulated in conjuction with an automatic thickness gauge which is built into the machine.

If rigid facings are required on one side, it is a simple matter to place these on the lower conveyor and to apply a flexible facing to the upper surface of the foam as before.

Rigid facings can be applied to both sides of the foam if one of them has sufficient flexibility to be bent round a fairly large radius. The mechanical side of this operation is obviously more complicated and requires further development but there are at present several machines with flexible facings operating commercially in various parts of the world.

In the transport industry, where the demand for refrigerated vehicles is continually increasing, injected rigid polyurethane foam has proved its superiority over the conventional insulants applied in slab form. By building the vehicle with the space between its inner and outer walls left empty and drilling a series of filling holes in the inner wall, polyurethane foam may be injected in a series of lifts until the space is completely filled (Fig. 5.14).

Fig. 5.14. Injection of polyurethane foam into vehicle wall cavities.

Alternatively, the outer wall only may be completed and a panel of plywood lined with polythene temporarily fitted to form the lower part of the inner wall. Polyurethane foam may then be poured between these two members and as soon as it has cured sufficiently (a matter of 15 minutes or so) the plywood panelling is moved upwards to form a mould for the next layer of foam, which bonds intimately with that already placed. By this means the whole of the vehicle may be lined with a jointless layer of insulant and an inner wall of any suitable material may be built in afterwards.

This ability to form a jointless insulation is of the greatest importance in these applications. It has been shown[2] that the ingress of moist air through joints in the structure and insulation of a refrigerated vehicle when travelling at speed can increase the heat losses by up to 36% over those measured when the vehicle is stationary. When tested under conditions equivalent to a road speed of 50 miles per hour, a conventionally insulated vehicle with 6 in of insulation has shown losses of 6160 Btu/h 100 deg F on account of sensible and latent heat leakage, compared with only 616 Btu/h 100 deg F for a vehicle insulated with only 4 in of injected polyurethane foam (Table 2). It may be noted that in addition to this marked reduction in heat loss, the internal capacity of the vehicle insulated with polyurethane foam had been increased by 160 ft^3 as a result of the reduced thickness of insulation necessary.

TABLE 2

HEAT GAIN TEST ON INSULATED TRAILER

Internal Temperature, 0°F. External Temperature, 100°F.
Applied Frontal Air Pressure Equivalent to 50 m.p.h. (Btu/h 100 deg F)

Source of heat gain	6 in Conventional insulation	4 in Urethane foam insulation
Transmission through insulation	5340	5285
Inward leakage of moist air (latent and sensible heat)	6160	616

Similar considerations apply to the insulation of road tankers, particularly those engaged in the transportation of liquefied gases such as carbon dioxide or ethylene. In such cases the inner barrel of the tanker is cradled in a series of rings made from polyurethane foam of relatively high density, having a high load-bearing capacity but still possessing extremely good insulation properties. The outer cladding is then built either by injection through holes after completion of the cladding or in a series of applications dispensed into the cavity as each layer of external cladding is built up.

Ships carrying refrigerated cargoes have for many years been insulated with rigid polyurethane foam applied by similar techniques. Apart from its excellent insulating properties, the impermeability of the foam to water and its ease of application in locations not easily accessible render it eminently suitable for this type of work and it is accepted by Lloyds Register of Shipping for this purpose.

The examples of rigid foam applications which have been described represent, not laboratory experiments, but actual commercial uses which have been established and have in many cases been in operation for several years. Polyurethane foam has, in fact, taken its place as one of the leading thermal insulants available today.

REFERENCES

1. J. M. Buist, *J. Cellular Plastics*, **1** (1965) 101.
2. *Plastic Foam Insulation Conference of The Society of the Plastics Industries, New York, April,* 1962, Heffner.

6

Thermal Insulation using Multiple Glazing

THOMAS A. MARKUS

Professor of Building Science, University of Strathclyde (*Great Britain*)

INTRODUCTION

In a general consideration of building insulation it is as well to put the window problem into its setting. There are three important facts to consider.

First, that in most buildings today the window is the weakest link in the chain of insulation; depending on the perimeter to area ratio of the plan it may account for 75% of the structural losses. Second, that if there are poorly fitting opening lights, the rate of air leakage heat loss may be several times as high as the structural loss itself. Third, that on account of the particular physical nature of glass, the window acts as a unique heat filter. It transmits heat by the normal mechanism familiar in the case of any thin sheet of homogeneous material; but it is also transparent to short-wave solar radiation (including, of course, that part of it lying in the visible portion of the spectrum) as a result of which it affects the heat balance of a room in a rapidly changing and highly fluctuating manner. Thus, on a bright, cold winter's day, it may act as a net source of gain while the sun shines upon it, but within a few hours it is a major source of heat loss. In order to understand its physical behaviour we must first take a brief glance at natural radiation and the transmission properties of glasses; this is important even though we are concerned primarily with heat loss and winter problems.

Finally, although in these days of glass-curtain-walled office building this should no longer be necessary, one must warn against over-simplified climatic notions, particularly that which divides British weather, for engineering design purposes, into "winter" and "summer". Most of our weather, apart from rare periods of continuous cold or continuous heat, is characteristically marginal. That is to say, the hourly and diurnal variations are large as compared with the differences between the seasonal means. In the case of the window, with its negligible thermal capacity and its transmittance of radiation, this is even more important to recognise than in the case of more substantial and opaque materials.

NATURAL RADIATION

The sun radiates energy almost as a black body at a temperature of about 6000°K. That is to say, outside the earth's atmosphere the energy distribution is an almost smooth, Planckian curve appropriate to a body at 6000°K (Fig. 6.1). The total energy content (known as "the solar constant") is in the region of 420 Btu/h ft^2. Water vapour, ozone, dust and other atmospheric constituents absorb various energy

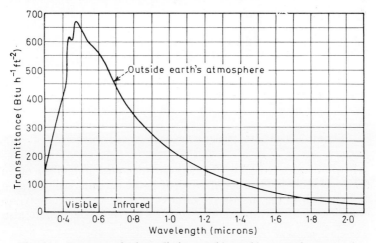

Fig. 6.1. Spectrum of solar radiation outside earth's atmosphere (moon).

bands until, at the earth's surface, something in the region of half the extra-terrestrial amount arrives. Figure 6.2 shows two spectra—one outside the earth's atmosphere and one for air mass 3, 19° solar altitude; in fact for each air mass corresponding to different altitudes of the sun above the horizon there is an appropriate incidence curve. The maximum intensity is, of course, at normal incidence to the solar beam; at other angles the intensity falls off as cos α where α is the angle of incidence. This has an important dual effect as far as vertical windows are concerned; the higher the sun is in the sky the greater is the normal incidence level, but the greater also is the cosine of the incident angle on a vertical surface. The two effects combine to give, typically, maximum radiation levels at solar altitudes in the 35° to 40° region—hence the higher intensities on E- and W-facing windows, in summer, than on S-facing ones, with their associated high altitudes.

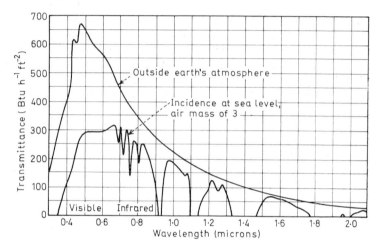

Fig. 6.2. Transmitted radiation through various air masses (moon).

It will be seen from Fig. 6.3 that the integrated area of the curve for direct radiation has about one-half in the u.v. and visible region and one-half in the short-wave infrared region, with, effectively, no radiation at wavelengths longer than 2·5 microns. The Figure also

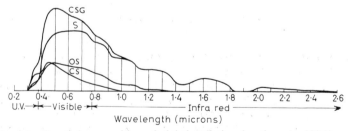

Fig. 6.3. Spectral composition of global radiation for clear sky (CSG), solar radiation (S), radiation from clear sky (CS), and from overcast sky (OS), all calculated for a horizontal plane with free horizon. The curves apply to a solar elevation of about 30° above the horizon. (Pleijel)

shows characteristic curves for the spectra of overcast and blue sky diffuse radiation. The intensity of the former adds about 15% to the level of direct radiation; it can, however, be much higher under hazy conditions typical of humid tropical atmospheres. Calculated and graphical techniques for the prediction of radiation intensities are available for most parts of the world and latitudes, including the British Isles.[1]

GLASS PROPERTIES

Figure 6.4 shows the typical transmission curve for a commercial window glass of $\frac{1}{4}$ in thickness for normally incident radiation. It will be seen that its transmission range corresponds very closely with the range of the incident natural radiation. At about 2·5 microns it virtually ceases. In most of the long-wave u.v., visible and short-wave infrared it is high, 70% or more, with a saddle-shaped dip

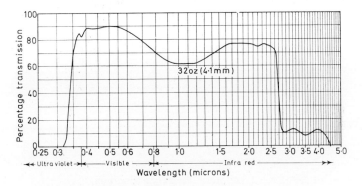

Fig. 6.4. Spectral transmission curve for 32-oz sheet glass.

characteristic of glasses having a fairly high iron content. Special heat-absorbing glasses having a deliberately enhanced iron content are also available, with a greatly increased drop in transmission in the short-wave infrared.

The transmission of radiation falls with increasing angle of incidence upon the glass; the drop is slight up to about 45° but it grows quite rapidly above that as Fig. 6.5 shows. Even at normal incidence roughly 4% is reflected at each surface of ordinary glass, giving a theoretical maximum of 92% transmission for a glass with zero absorption. Figure 6.5 also shows that the sum of the three fractions of the incident radiation—the transmitted, the absorbed and the reflected—is always equal to the original incident radiation (*i.e.*, 1·0), irrespective of the incident angle, but that the *ratios* between the fractions alter continuously. This tripartite division is only a function of the type of glass and spectral composition of the incident radiation, and is independent of the intensity of the incident radiation or the air temperature and air flow conditions at the two surfaces. Thus, if the normal incidence transmission for a glass is known, and its refractive index (which determines the quantity reflected) also, the full tripartite diagram can be produced.

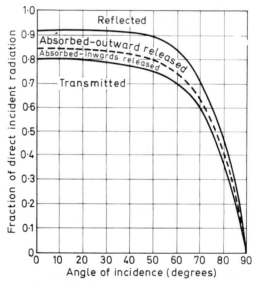

Fig. 6.5. Transmission, reflection and absorption of ¼-in plate glass, together with indication of inwards- and outwards-released absorbed energy.

That portion of the energy which is absorbed in the glass has to be dissipated somehow. In fact it raises the glass temperature and, when steady state is reached, is lost to both sides by convection and radiation. The sub-division of the absorbed component on Fig. 6.5 into "in" and "out" shows, for a particular set of conditions, the relative amount dissipated to the two sides.

The distinction between this heat dissipated by convection and *long-wave* radiation and the directly transmitted energy (short-wave) is an important one for predicting the thermal response of the room to solar radiation. The absorbed heat released into the room from the glass can be taken, approximately, as being a sensible heat gain whereas the short-wave radiation transmitted suffers both decrement and time lag before affecting room temperatures since it first has to be absorbed by the floor, walls and contents and then gradually released. Under summer conditions this can be the determining factor in the sizing of cooling plant and even, in extreme cases, in the decision on whether a cooling plant is necessary. It can be shown that over short periods, of a few hours, the drawing of a dark, absorbent blind across the inside of a window can result in a more substantial rise in the room temperature than that occurring with the sun shining through the unshaded glass.

In this paper we are chiefly concerned with insulation and hence, by implication, with cold weather conditions.

The chief effect of absorbed radiation in the glass is to alter the surface temperatures of the glass and hence the heat flow between the surfaces in such a way that the quantity of heat transferred is no longer directly proportional to the difference in air temperature on the two sides, as is usually assumed in the standard thermal transmittance (U value) calculation. This standard U value can only be strictly accurate under conditions when the surroundings are at air temperature. This is, of course, in practice, very rare; it is, however, a useful basic way of comparing materials and constructions, in which their U value is dependent only on the wind speed in the vicinity of the extreme surfaces, their emissivities and the conductivities of intermediate materials and conductances of intermediate airspaces. It is *independent* of the rate of heat flow through the construction and of the amount of incident radiation. Following Loudon's argument in connection with opaque materials,[2] and with a slight alteration to his terminology, we shall call this the *effective transmittance*—U_e. We must therefore first consider the factors which determine effective transmittances for various types of glazing systems.

HEAT FLOW THROUGH GLASS

Effective transmittance (U_e) of glazing

Under conditions of proportionality between the air temperatures on the two sides of the glazing and heat flow (H), *i.e.*, when

$$H = \frac{T_{ai} - T_{ao}}{R_e} \qquad (1)$$

where T_{ai} = internal air temperature,
T_{ao} = external air temperature,
and R_e = effective resistance of the glazing,

a typical temperature gradient is established in the glazing as shown in Fig. 6.6.

The total effective resistance R_e is the sum of the extreme surface resistances, R_{si} on the inside and R_{se} on the outside, the resistance of any layers of materials, e.g., glass, R_g, and of any airspaces R_a. Thus

$$R_e = R_{si} + R_{se} + R_g + R_{a1} + R_{a2}, \text{ etc.} \qquad (2)$$

From eqn. (2), the effective transmittance is the reciprocal of the effective resistance, i.e.,

$$U_e = \frac{1}{R_{si} + R_{so} + R_g + R_{a1} + R_{a2}}, \text{ etc.} \qquad (3)$$

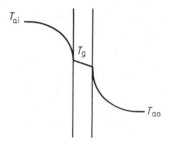

Fig. 6.6. Temperature gradient through single glass when surroundings are at air temperature.

Surface resistances

There is insufficient space here to enter into detailed consideration of values of R_{si} and R_{se}; reference should be made to a standard work on heat transfer.

The internal surface resistance is made up of two constituents; the resistance to convection and the resistance to radiation. It is the reciprocal of the sum of the two coefficients, thus

$$R_{si} = \frac{1}{h_c + Eh_r} \qquad (4)$$

where h_c = the convection coefficient, h_r = the radiation coefficient and E = the emissivity.

The conventional value for the inside surface resistance R_{si} is 0·7 ft² h deg F/Btu. This makes certain assumptions about wind speed and the Mean Radiant Temperature of the room surfaces which are described in the literature.

The outside surface resistance R_{se} for surfaces of normal emissivity varies according to wind speed. Values are given in the IHVE Guide for various degrees of exposure to winds and orientations, that for "normal" exposure to wind being conventionally taken as 0·3 in the same units as before. The Guide gives this value for N, NW, NE and E orientations; as will be seen, it is not possible to build into such standard values the effect of radiation (*i.e.*, orientation).

Glass resistance

The conductivity for glass is about 7·3 Btu/h ft² in. In the small thicknesses normally used its resistance is therefore so low that in most calculations it can be ignored; for instance in the case of ¼ in-thick glass

$$R_g = \tfrac{1}{4} \times \frac{1}{7·3} = 0·03 \text{ ft}^2 \text{ h deg F/Btu.}$$

Airspace resistance

The heat flow across an air space depends on:

(a) Orientation; in a vertical airspace there is a rising stream of air on the warm side and a dropping one on the cold side. In a horizontal space there are laminar layers of air of decreasing temperatures between the warm side and the cold; without local turbulent effects convection currents and consequent convection exchange would be suppressed. In a sloping air space the state is at some intermediate condition between that in a vertical and that in a horizontal space, according to the angle of the slope.

(b) The temperature difference between the two sides; the higher this is, the greater the heat exchange.

(c) The actual temperature; the radiation and the convection coefficients are both temperature dependent, increasing with increasing temperature.

(d) The height:width ratio of the airspace; it is only when this is sufficiently large (say greater than 20:1) that the idealised conditions assumed in this discussion are true.

The heat exchange takes place by both radiation and convection/conduction.

The radiation exchange depends only on absolute temperature and temperature difference; it is independent of airspace width.

The convection/conduction coefficient depends not only on temperature difference and airspace slope, but also on airspace thickness. Figure 6.7 from Robinson, Powlitch and Dill[3] shows the variations of the coefficient with thickness for vertical spaces at five temperature differences. It will be seen that at narrow spaces the transfer is by conduction only (dotted line), decreasing in an exponential manner with increasing thickness of air—e.g., halving for every doubling of air mass. But beyond a critical distance, which depends on the temperature difference, transfer takes place primarily by convection and expected decreases due to increased air mass do not actually take place. Below this critical distance the two layers of warm and

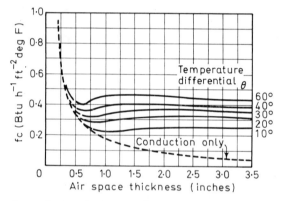

Fig. 6.7. Convection/conduction coefficient for vertical air spaces (after Robinson, Powlitch and Dill).

cold air are sufficiently close to each other for frictional forces to prevent free sliding of the layers past each other, and thus suppress free convection. Above this distance this is no longer true. For temperature differences of about 10 deg F the minimum exchange takes place at about $\frac{3}{4}$ in; it rises above this and drops to a second minimum at a much larger (normally impracticable) distance. For 30 deg F difference the first minimum is at a distance of $\frac{3}{8}$ in.

The reciprocal of the sum of the radiation and convection/conduction coefficients gives the total airspace resistance, often taken for simplified calculations as about $1 \cdot 0$ ft^2 h deg F/Btu.

If only T_{ai} and T_{ao} are known and an accurate calculation is needed of the appropriate airspace resistance, an initial approximation is made, from which appropriate surface temperatures for the airspace boundaries are obtained, which in turn yield a more accurate value for airspace resistance and so on; thus successive approximations can be used until the improvement is insignificant.

Glass surface temperatures

The temperature at any point of a single- or multiple-glazed structure can be found quite simply from the fact that, in the steady state, the heat flow across each point must be equal and therefore the temperature drop will be a function of resistance to the flow of heat at that point—a large drop for a large resistance, and a small one for a small resistance.

Thus the heat flow H at any surface equals the temperature difference between the surface and the air divided by the surface resistance; for instance, at the inside surface

$$H = \frac{T_{ai} - T_g}{R_{si}} \qquad (5A)$$

from which it follows that

$$T_g = T_{ai} - HR_{si} \qquad (5B)$$

H, of course, being computable for any structure the R_e of which is known, from eqn. (1).

Effective transmittance, U_e

From eqn. (3) we are now in a position to compute, for any single- or multiple-glazing system, the effective transmittance U_e. Taking internal and external surface resistances at 0·7 and 0·3 respectively, and the airspace resistance at 1·0, we obtain the conventional values of 1·0 and 0·5 for single- and double-glazing respectively (ignoring the small glass resistance), all in the conventional units. Of course, for proper comparison and calculation, all the components have to be accurately computed.

Effect of radiation on glass

At the outer surface of glass there may be a complex radiation exchange—short-wave radiation gain from the sun, sky and reflecting surfaces; and long-wave radiation loss to the ground and sky. Loudon[2] summarises available data, and his method of dealing with this radiation exchange is followed here.

The way this radiation exchange is accounted for is to use the effective transmittance (U_e) value but to include, also, a term which accounts for the net radiation effects, yielding a new value U_a which is the actual air-to-air transmittance under the unique conditions of temperature, radiation and wind speed in question. On north-facing facades, and in winter, the differences between U_e and U_a may be sufficiently small for it to be valid to use computed U_e values for design purposes. However, for other seasons and orientations the differences cannot be ignored and values of U_a should be used which take actual radiation conditions into account as shown here.

At the outer face of the glazing the net heat gain is the difference between the heat gain by short-wave solar radiation and the heat loss by long-wave radiation to sky and ground.
Thus

$$H_{re} = \alpha I_s - E I_l$$

where H_{re} = the excess heat gain by radiation at the outside surface; it may be negative in midwinter
I_s = incoming solar radiation intensity (Btu/ft² h)
I_l = long-wave radiation loss from a black surface at air temperature with the same orientation as that of the surface in question (Btu/ft² h)
and α = absorptivity of glass for solar radiation.

Conventionally, the actual outside surface resistance (R_{so}) is given by

$$R_{so} = \frac{1}{H}(T_g - T_{ao}) \tag{6A}$$

but,

$$H = \frac{1}{R_{se}}(T_g - T_{ao}) - H_{re}. \tag{6B}$$

Multiplying eqn. (6A) through by R_{se} and dividing by H we get

$$R_{se} = \frac{1}{H}(T_g - T_{ao}) - \frac{H_{re} R_{se}}{H}. \tag{7A}$$

The first term in eqn. (7A) is the same as the conventional expression for R_{so} as given in eqn. (6A). Therefore

$$R_{se} = R_{so} - \frac{H_{re} R_{se}}{H}. \quad \text{(N.B. In the absence of the radiation term } R_{se} = R_{so}.\text{)} \tag{7B}$$

Hence
$$R_{so} = R_{se} + \frac{H_{re}R_{se}}{H} \quad (7C)$$

$$= R_{se}\left(1 + \frac{H_{re}}{H}\right). \quad (8)$$

H_{re} could be rewritten as $\dfrac{R_{se}H_{re}}{R_{se}}$ by multiplying top and bottom by R_{se}.

Then
$$H = \left(\frac{T_a - T_g}{R_{si}}\right) = \left(\frac{T_g + T_{ao} - R_{se}H_{re}}{R_{se}}\right) \quad (9)$$

and adding the top and bottom lines of eqn. (9) (ignoring the glass resistance and temperature drop across its two faces) we obtain on the top line the total air-to-air temperature drop less a radiation term and on the bottom line the total effective resistance. Hence,

$$H = \frac{T_{ai} - T_{ao}}{R_e} - \frac{R_{se}H_{re}}{R_e}. \quad (10)$$

Since $\dfrac{1}{R_e} = U_e$,

$$H = U_e(T_{ai} - T_{ao}) - U_e(R_{se}H_{re}). \quad (11)$$

The first term in eqn. (10) is the normal air-to-air heat loss term; the second is the effective radiation gain. It is seen that the amount of the total radiation gain H_{re} which is admitted to the room is determined by the ratio of the resistance to one side—the outside—of the absorbing layer to the total effective resistance. This is a very important generalised finding, since it means that the closer the absorbing layer lies to the outside of the glazing system, and hence the smaller the proportion of the total resistance which lies to the outside, the lower the amount of admitted heat. It explains why, for prevention of excessive heat gain, in a double-glazing system, it is better to have the heat-absorbing layer as the outside one, and why an external blind is more effective than one on the room side; in the latter case virtually the whole of the resistance of the system lies to the outside of the absorbing layer (the blind) and hence the radiation gain approaches 100%.

To conclude this section, let us take two simple examples illustrated in Figs. 6.8 and 6.9.

74   THERMAL INSULATION

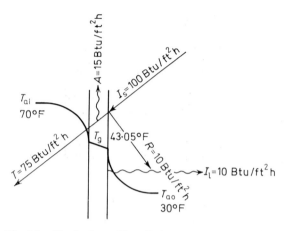

Fig. 6.8. Single glass with radiation–temperature gradient.

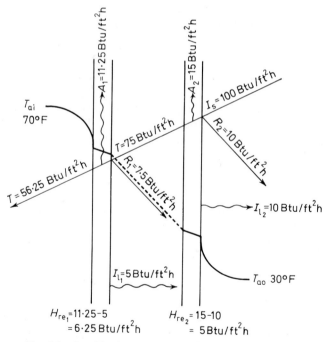

Fig. 6.9. Double glass with radiation–temperature gradient.

MULTIPLE GLAZING

Figure 6.8 shows a single glass, with incident short-wave radiation $I_s = 100$ Btu/ft² h. The glass has an absorptivity $\alpha = 0.15$ and the outgoing long-wave radiation $I_1 = 10$ Btu/ft² h. $T_{ai} = 70°F$ and $T_{ao} = 30°F$. $R_{se} = 0.3$ ft²h deg. F/Btu.

Therefore $H_{re} = 5$ Btu/ft² (the net quantity of radiation gain). From eqn. (11), and taking U_e for single glass = 1.0 Btu/ft² h deg F,

$$H = 1(70 - 30) - 1(0.3 \times 5) \text{ Btu/ft}^2 \text{ h}$$
$$= 40 - 1.5$$
$$= 38.5 \text{ Btu/ft}^2 \text{ h,}$$

instead of 40 which would have been given using the conventional value of U_e alone without making a radiation allowance.

Putting $\quad 38.5 = \dfrac{T_{ai} - T_g}{R_{si}}$ (taking $R_{si} = 0.7$)

gives $T_g = 43.05°F$ instead of $42°F$ when $H = 40$ Btu/ft² h deg F.

In fact the value of R_{si} is a function of T_g, so with a change in the latter a new value of the former should be used. Normally, however, the difference is insignificant; where conditions are such that this is not true, a series of successive approximations of T_g and R_{si} are made until no significant improvement is obtained.

In the case of the double-glass system shown in Fig. 6.9 the conditions and assumptions are the same as in the previous example. In addition it is assumed that both glasses have the same transmission and absorption properties, and that as a result of the higher temperature of the inner glass as compared with that of the outer one the radiation from the inner to the outer is equivalent to 5.0 Btu/ft² h.

Therefore $H = \left(\dfrac{T_{ai} - T_{ao}}{R_e}\right) - \left(\dfrac{H_{re1} R_a + R_{g2} + R_{se}}{R_e}\right)$

$$- \left(H_{re2} \times \dfrac{R_{se}}{R_e}\right)$$

$$= \text{(say)} \left(\dfrac{70 - 30}{2.0}\right) - 6.25 \dfrac{(1 + \ldots + 0.3)}{2.0} - \dfrac{5(0.3)}{2.0}$$

$$= 15.19 \text{ Btu/ft}^2 \text{ h.}$$

In the absence of radiation gains, this would have been 20.0 Btu/ft² h. Glass-surface temperatures can now be computed throughout the glazing system in the same way as before.

It is important to notice that none of the calculations takes any account of what happens to the *transmitted* short-wave radiation. This is a separate problem, closely bound up with the nature of the room surfaces and contents which absorb it, and requiring calculation of decrement and time lag; it is outside the scope of this chapter.

Frame effects

Heat loss through the frame material of a window may be substantially larger than that through the glass—the so-called "cold bridge" effect. Table 1 shows typical values for frame factors by which glass U values should be multiplied to obtain overall window U values for various frame materials and ratios of frame:glass. It will be seen, for instance, that in the case of an aluminium-framed double-glazing system, the increase in transmittance is 20% for frame:glass ratios of 1:4. Special precautions can be taken to break the "bridge" by inserts of low conductivity, such as nylon, but these are expensive frames and not effective if there are even occasional through rivets or bolts, which act as effective channels for heat conduction, largely nullifying the effects of any continuous breaks.

TABLE 1
FRAME FACTORS*

	Single glass		Double glass	
	% Glass	Factor	% Glass	Factor
Glass	100	1·00	100	1·00
Wood frame	80	0·90	80	0·95
Wood frame	60	0·80	60	0·85
Steel frame	80	1·00	80	1·20
Aluminium frame	80	1·10	80	1·30

* Based on values from the ASHRAE Guide.

Curtains and blinds

Billington[4] reports a decrease in transmittance from 0·59 to 0·48 resulting from the use of light curtains on timber-framed single windows, with a further decrease to 0·43 if the curtains are heavy and well fitting. Pleijel[5] found significant drops in heat loss through windows fitted with fabric blinds having a metallised surface facing the glass, effectively increasing the airspace resistance by a reflective and low emissivity surface on one side. The use of curtains at night, although almost universal in most buildings during the heating season, is rarely allowed for in window heat loss calculations, with consequent excessive losses predicted.

CONDENSATION

One of the reasons for using insulating glazing is to prevent or reduce the incidence of condensation on glass. Condensation will occur whenever the surface temperature of the glass drops to the dewpoint temperature of the air next to it, or below. The dewpoint is that temperature at which air, containing a given amount of moisture, has a 100% relative humidity—*i.e.*, is saturated. Any rise in air temperature, without a change in the moisture content, results in a lowering of relative humidity, and any drop results in condensation—or frost, if the dewpoint is at or below freezing point.

The prevention of condensation is generally important for one or both of two reasons: to prevent damage from collections of liquid water and to preserve clear vision through windows. In addition, in certain industrial and scientific buildings where close humidity control is needed, it may be necessary to prevent window glass "condensing out" moisture which is needed in the atmosphere.

We have already seen how, under a given set of conditions, the surface temperature of glass can be computed, to check whether it is at or below dewpoint. This check can be carried out either for the inside surface of single glass or the innermost surface of a multiple system, or for the surface of the boundary of one of the airspaces. Four independent factors are involved: (i) the thermal resistance of the glazing system between the surface under consideration and the outside, (ii) the relative humidity of the room or airspace air, (iii) the room or airspace temperature and (iv) the outside air temperature. If any three of these are known the fourth can be predicted. This prediction is simplified by the use of such a graph as that shown in Fig. 6.10.

Condensation within airspaces

Condensation will occur on the bounding glass surface of an airspace whenever its temperature reaches the dewpoint of the enclosed air.

Unless the airspace is permanently hermetically sealed, water vapour will be able to enter it by one or both of two independent mechanisms. The first is by diffusion of vapour from the zone of higher vapour pressure to the zone of lower. This is independent of the total (atmospheric) pressure of the air/water mixture and can take place even when the pressures are the same but the mixtures are composed of different proportions of air and water. Only molecular sized apertures are needed for the transfer. The second is caused by differences in total atmospheric pressure (*e.g.*, due to wind, convection or turbulence) as a result of which there will be a transfer of air/water mixture (and with it, of course, of water vapour) from the zone of higher pressure to the zone of lower.

In practice one can use a permanently hermetically sealed double or multiple glazing unit, in which the air is dried during the manufacture of the unit by the sealing-in of an adequate quantity of

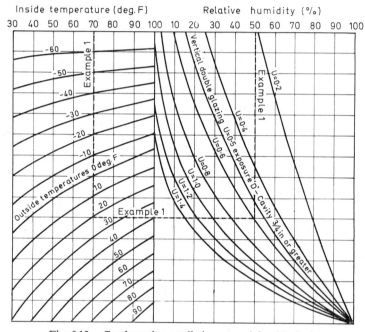

Fig. 6.10. Condensation prediction curve (after Markus).

desiccant or is dried before being enclosed. Alternatively, one or more of the following four precautions may be taken to minimise the incidence and consequences of condensation:

(a) Seal the airspace effectively from the moist side, usually the room side. This may mean the use of special compression gaskets. Also it is important for any timber facing the airspace to be adequately sealed, otherwise, especially under sunlit conditions, water vapour will be released from it and cause temporary condensation.

(b) Allow adequate breathing or ventilation of the airspace to the drier side, usually the outside. This will allow both diffusion of vapour from the space to the drier outside air and ventilation due to any pressure differences that may be set up. Even quite small gaps are effective. Nycander[6] found that a gap of 1 mm in the case of a coupled sash double window was adequate to prevent condensation without any significant reduction in the resistance of the space to heat flow. Pratt[7] also investigated the effect of ventilating cavities in roofs and found little loss of insulation at air speeds necessary to prevent condensation on the underside of sheeting.

Openings should, if possible, be fitted with filters to prevent dust and insects from entering; loosely packed glass fibres or sintered metal pellets can be effective.

(c) Allow any condensate to be collected and drain out through weep holes and tubes.

(d) Enclose a cartridge of desiccant. In this case it is important to seal the airspace to *both* sides, to ensure that as little moisture as possible enters. In any event the desiccant will have a limited life and should be in a container that can easily be removed from the room side and replaced after reactivation or renewal of the material.

TYPES OF DOUBLE AND MULTIPLE GLAZING

Most standard works on construction give details of various systems. It is sufficient here for the main types to be listed, with brief comments on their properties.

Hermetically sealed units

Many manufacturers offer proprietary makes of various types of glass made into these units. Airspaces vary from $\frac{1}{8}$ in to $\frac{1}{2}$ in and the seals are either various combinations of metal, plastics and adhesive, or, more recently, fused glass in which the panes are joined to glass edge seals with no other material involved. Unless special precautions are taken (*e.g.*, the provision of a balloon-type expansion valve as in aircraft windows) such units are liable to assume varying curvatures on account of changes of atmospheric pressure and temperature; normally this is unimportant, but it may matter in certain situations where undistorted reflections are particularly needed.

Provided that the airspace is adequately dried and sealed, these units can avoid condensation completely, and involve only as much cleaning as is needed for a piece of single glass—the two extreme surfaces. They tend to be relatively expensive.

Double patent glazing

This is an adaptation of the traditional "puttyless" glazing system often used in factory roofs, in which the two pieces are glazed into the patent glazing bars and separated by a metal or other spacer. The glazing here is not hermetically sealed and therefore both condensation and dirt will occur in the space. It may be necessary to dismantle the glazing for cleaning every few years, but the initial cost is relatively low.

Double windows

A great variety of timber and metal coupled sashes, double casements and sliding windows are available; none is hermetically sealed but there is generally provision for ventilation or breathing to the outside and for easy separation for cleaning. Costs are relatively high.

"Do-it-yourself" systems

Recently there has been a rapid growth in the number and types of systems in which an additional pane of glass can be fitted to existing, single windows. Frames are in timber, plastics and metal and all are relatively cheap. Where they are fitted over existing poorly fitting opening lights they achieve the additional benefit of reducing air leakage into rooms.

Extrusions

High-pressure gaskets of the type originally used in the transport industry are now available in which "E"-shaped or other rubber or plastics sections accept two panes of glass which are then held under high edge pressure by the insertion of a "zip-in" pressure strip. Alternatively the gasket may be compressed by mechanical means in the construction of the frame.

ECONOMICS OF DOUBLE AND MULTIPLE GLAZING

One of the critical factors in the decision whether or not to use insulating glazing is the cost benefit. Whilst the calculation of costs is relatively simple, it is limited to the cost of glazing, cleaning and maintenance, the possible capital savings on heating plant due to reduced peak load and possible savings on fuel energy resulting from a reduction in average heat loss. The technique is illustrated below. There are some general points to be made, however.

The first is that in calculating the difference in seasonal energy consumption between single and double glazing, for instance, the calculations should be based on equivalent thermal *comfort* and not on equal air temperatures. Using, for instance, the Bedford Equivalent Temperature scale, allowance can be made for the fact that the inner surface of single glass will be substantially colder than that of double glass under similar environmental conditions. This will alter the Mean Radiant Temperature, and MRT is, as a rough approximation, the factor in thermal comfort which takes into account the radiation exchange of the human body with its surroundings. Moreover, since colder air will be dropping down the surface of the single glass, it may be necessary to calculate the E.T. for a higher wind speed. In this way, under typical conditions for a person sitting near a large window, one may find that the appropriate air temperature for a given degree of thermal comfort is 2–3 deg F higher for single than for double glass; this, over a heating season, can result in a significant difference in energy compared with a calculation based on equal temperatures.

The second point is that certain benefits of insulating glazing are impossible to cost in an ordinary way, notably improved sound insulation (if the airspace is at least 4 in, preferably larger) and absence of condensation. The only way these could be included is by assigning to these benefits a cost saving based on a value analysis system, which would include ranking or other comparative scaling with other environmental and building factors.

For the straightforward cost exercise it is necessary to reduce all capital, recurring and annual costs to a single figure for comparison purposes. The cost of glazing and windows, and the cost of heating plant (determined by *peak* demand) are capital sums. The cost of reglazing, painting, renewal of desiccant and so on, may be recurring costs, *e.g.*, every 5 years. The cost of cleaning and fuel energy is an annual one. Quite a convenient way of treating all these costs is by considering their "present worth". In other words, by making proper allowance for the life of the building, and expected interest on invested capital, one can obtain the equivalent single, capital sum which today is worth the same amount as a series of annual or recurring sums in the future.

To find the Present Worth (P) of a uniform annual payment A over a period of y years at $i\%$ interest,

$$P = A\frac{(1 + i)^y - 1}{(1 + i)^y}.$$

Thus an annual expenditure on fuel of £5 for 20 years at 5% interest has a Present Worth of £62·35. This is the amount of capital it would be worth spending today, say on extra insulation, to avoid paying this annual sum. Any capital expenditure greater than this represents a loss.

Similarly a single or recurring capital sum F in the future can be discounted to its Present Worth by

$$P = F\frac{1}{(1 + i)^y}.$$

The Present Worth sums are added to actual initial expenditure to make valid comparisons between one alternative and another on a cost-in-use basis over the life of the building.

REFERENCES

1. P. Petheridge, *Sunpath Diagrams and Overlays for Solar Heat Gain Calculations.* Building Research Station, D.S.I.R., *Building Research Current Papers, Research Series No. 39 and Supplement,* 1965.
2. A. G. Loudon, *J. Inst. Heating and Ventilating Engineers,* Nov. (1963) 273.
3. H. E. Robinson, F. J. Powlitch and R. S. Dill, *Research Paper No.* 32, Housing and Home Finance Agency, Washington, 1954.
4. N. S. Billington, *Thermal Properties of Buildings,* Macmillan, London, 1952, p. 34.
5. G. Pleijel, *Byggmastaren,* **36** (1957) 201.
6. P. Nycander, *Varmeisolering och Kondensering hos Fonster: Inverkan av Glasavstand och Ventilation Mellan Glasen,* Sverige Statens Provningsanstalt, Meddelande 96, Stockholm, 1946.
7. A. W. Pratt, *National Building Studies Research Paper No.* 23, D.S.I.R., London, 1958.
8. *Glass and Windows Bulletin,* No. 1, Pilkington Bros. Ltd., p. 6.

7

Dual Purpose Materials
Thermal Insulation and Sound Absorption

J. LAWRIE

Structural Insulation Association, London, W.1 (Great Britain)

INTRODUCTION

Thermal insulation has been fairly widely studied and discussed during the last twenty years and although in the building trade the adopted standards, as evidenced by the Building Regulations of 1965, are still below the minima recommended by the Burt Committee twenty years ago there is a general awareness of the subject in both industrial and social circles.

The same is not the case with sound insulation around which there is a general cloak of mystery, and even the most practical aspects are only vaguely understood by anyone other than the trained acoustics engineer. Sound insulation, sponsored by the Wilson report and now gaining a first mention in the Building Regulations, is at the same stage of general interest as thermal insulation had reached twenty years ago.

The social requirements of the next decade will most probably bring unwanted sound to the forefront of mass consciousness as has been done with domestic heating over the past decade. The present knowledge will be assimilated over a wider scale and the air of mystery removed.

This chapter sets out to discuss unwanted sound in practical terms, to show how the types of sound and the transmission paths involved affect the methods of treatment, to indicate the limitations of these methods of treatment and to show how the requirements can be similar to those of thermal insulation.

Materials which can perform the dual function of thermal insulation and sound insulation are discussed fully as are their limitations in this sense. Applications over a wide field in the building, heating and transport industries are covered.

THERMAL INSULATION AND SOUND INSULATION

The basic principle of thermal insulation—the erection of a barrier against the transmission of heat—is well known. Extra clothes are worn in wintry weather; these of course do not generate heat, but simply help to retain the natural warmth of the body. Heating equipment is covered in and pipes are wrapped in the interests of economy and the prevention of damage. Thermal insulation may be said to be a part of general life.

Sound insulation, however, is in a very different position. Only in the last few years has the level of unwanted sound reached the point where it has become the subject of general comment and while this comment may be widely publicised, knowledge of the possible remedial measures is so low that even Governments have been known to finance sound insulation projects which have virtually no chance of success.

A clear understanding of the nature of sound and its transmission is not easy to acquire, but it is desirable that at least some working knowledge of them should be gained before remedial measures are considered.

Sound is a form of energy which is transmitted in all directions, in waves of equal intensity. These waves will travel in straight lines, losing energy as they go until they finally die out. If they meet an intervening object they will be absorbed or reflected according to the nature of the object. Sound can be produced at different frequencies, as for instance in the notes on a musical scale, but frequency is not connected with loudness. We can play a top C or a middle A with equal loudness, but each note has its own frequency.

Unwanted sound can be defined either in terms of degrees of loudness objectionable to the human ear, or in terms of frequencies equally objectionable. Unwanted sound is usually called "noise" but the word is somewhat misleading in view of its general identification with loudness.

It has already been said that when a sound wave impinges on an object it will be absorbed or reflected according to the nature of the object. If the object has a hard surface, reflection will occur and if soft, absorption; but more often some combination of reflection and absorption is probable. This arises from the fact that sound energy is easily converted into mechanical or heat energy and when a sound impinges on the hard surface of a thin lightweight material some of its energy is used to vibrate the lightweight material and only part of the sound is reflected. Similarly, when sound impinges on the soft but thin surface of a thick, dense material only a proportion of the sound will be converted into heat energy and the remainder will be reflected.

Thus only thick, dense, hard-surfaced materials can be considered truly reflective and thick, light, soft-surfaced materials truly absorptive.

In the case of thin hard-surfaced lightweight material which reflects and also absorbs by its own vibration, the reflection can be reduced and the absorption increased by the incorporation of a facing which is thick, soft and lightweight, but if the base material is vibrating sound will be transmitted to the other side.

Similarly, with soft thin-surfaced dense material, absorption may be increased by thinning the dense material until it vibrates under the effect of the sound impingement, but transmission will occur, on the opposite side and little will have been done to assist in reducing sound on the impingement side.

The reverse applies for the thick soft material which is highly absorbent on the impingement side, but may offer little resistance to the passage of sound through it to the other side.

An added complication arises from the fact that dense materials do not behave exactly according to weight alone. The form in which the material is used also has an effect. If, for example, sound transmission measurements are carried out, first through a cardboard plate 2 inches thick and then through a panel consisting of four $\frac{1}{2}$-inch-thick plates placed one inch apart, it will be found that the value obtained in the first instance is much greater than that with the four-plate assembly although the total weight is the same in each case. This lamination principle is used extensively in dealing with sound insulation problems.

Again, thin materials can be made to act in a manner contrary to their normal behaviour if they can be backed with firm but lightweight materials capable of damping out the vibrations of the thin material and not transmitting these vibrations through themselves to the other side.

The complexity resulting from the inter-relationship of these factors is great and yet all these factors are present in relation to the simplest sound problem. It is necessary therefore to consider each problem by itself, to identify the principal factors involved and then to decide on the best method of treatment.

Sound, although in the main airborne, more often than not has or develops a structure-borne component. It can also be basically structure-borne when it is the result of impact on the structure, as for instance in the case of a footfall. The nature of the sound and its path will govern the method of control used; for airborne sound massive construction and absorption will be in order, while structure-borne sound will call for dissociation and isolating membrane technique. It is necessary therefore to be quite clear about these factors also before treatment can be considered.

The last point to be dealt with is the nature of the problem itself. Since unwanted sound is clearly a subjective matter we must identify the reason for the objection (which may not be obvious), check as to whether the objection is uniform in relation to time, ascertain the degree of objection and hence decide whether treatment is necessary or indeed possible. Only then can the methods of treatment available be considered.

The factors affecting the solution of any sound problem have been discussed at some length, but this has been done only because many of these aspects are not obvious and great mistakes can occur if a clear picture is not obtained.

A series of problems common within their industries will now be taken, to show how the principles covered can be applied in practice. The cases cited will be those where thermal insulation and sound insulation requirements can be met with one application.

DUAL PURPOSE APPLICATIONS

Building

There are many thermal/sound problems encountered in the sphere of building, but only a few can be dealt with in this chapter.

In a block of flats, for instance, it could happen that a quiet elderly couple have very noisy neighbours and also perhaps that heavy-footed people occupy the rooms overhead. If in the construction of these premises due consideration has not been given to problems of sound, conditions for these peace-loving people could be very trying. If, at the same time, no proper account has been taken of the thermal requirements, conditions would be even less tolerable.

Several straightforward precautions should have been taken at the design stage (Figs. 7.1 and 7.2). The party wall between the flats could have been constructed of two leaves of insulating concrete giving good thermal insulation and better sound insulation than eleven-inch cavity brickwork.

The floor/ceiling construction, basically of timber, should have incorporated a floating floor, the resilient membrane between the floating floor and the joists being a glass or mineral wool quilt. This construction when used in conjunction with sand pugging at 17 lb/ft^2 not only reduces airborne and impact sound transmission to a comfortable level, but also reduces heat flow by more than 50%.

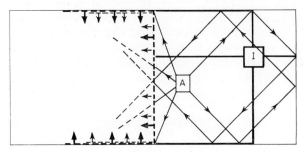

Fig. 7.1. Flats—ground plan.

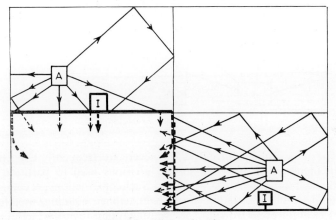

Fig. 7.2. Flats—elevation.

Unfortunately, as is often the case with problems of this type, the solution is easily incorporated in the basic design, but almost impossible to add at a later stage.

Not all problems, however, are incapable of solution after the building is constructed.

Consider a factory with thin outside walls, built prior to the application of the Thermal Insulation (Industrial Buildings) Act,

88   THERMAL INSULATION

1957. From a thermal viewpoint this building is very inadequate and if it houses a noisy process, conditions for its workers must be well-nigh unbearable (Fig. 7.3).

A roof and wall lining of insulated fibreboard will reduce both heat loss and reverberation making conditions inside the factory much more pleasant. A further improvement on both counts can be obtained if a glass or mineral wool mat is added behind the fibreboard or if the same backing is used behind a perforated plasterboard or hardboard lining. This type of treatment can, of course, apply equally well to any type of structure with a similar problem.

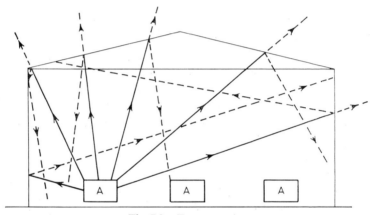

Fig. 7.3.   Factory section.

The most difficult problems, however, are frequently concerned with lightweight, often temporary, divisions used to partition off individual rooms within larger spaces. Such constructions are common in modern offices or in temporary light-skinned buildings used for a variety of purposes from doctors' waiting rooms to airport terminals. These problems usually relate to the difficulties of operation when an area of noise is close to an area requiring quiet or where an area of high sound security is next to one of low security.

An example of the combination of these problems can be visualised in a company director's office separated from his secretary's office on the one hand and the general office on the other by lightweight partitions. During periods of noisy activity outside his room the director finds the sound distracting; when there is quiet, he is conscious that conversations within his office can be heard clearly outside (Figs. 7.4 and 7.5). The requirements are for peace *and* security, and in a case of this sort there is no one solution. For satisfactory results a series of measures has to be employed.

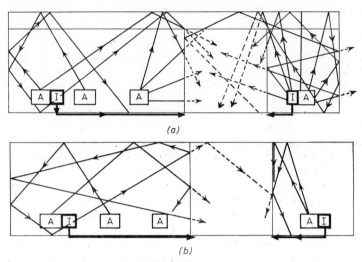

Fig. 7.4. Offices (a) with separating walls short of ceiling—elevation; (b) with separating walls up to ceiling—elevation.

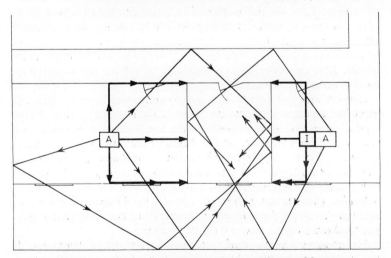

Fig. 7.5. Windows and doors and floor/wall vibration—plan.

The first point to check perhaps is to see that the typewriters are placed on felt pads or fitted with shock-absorbent mountings. The basic problem then is to ascertain the various sound paths involved. The most obvious path is, of course, through the intervening partitions, and so the next step would probably be to make sure that the tables or desks on which the typewriters and telephones stand are out of contact with the walls. If there is a suspended ceiling in the offices, it must be found out if the partitioning is carried up to the soffit behind the suspended ceiling. If not, there is a concealed path for the sound to follow. In the same way sound can pass through the windows and doors.

After these matters have been attended to there remain the stiffening of the partitioning itself and the reduction of the overall sound level in the offices.

Treatment to the partition would in theory involve a considerable increase in mass and the provision of a hard reflecting surface. In practice, however, this is unlikely to be possible, and a more acceptable solution would be to face the partition with a thin dense sheet, e.g., 1-inch plasterboard, backed with a resilient layer of glass or mineral wool of a density of 2–3 lb/ft^3 for the finer fibre materials, and 6–8 lb/ft^3 for coarser varieties. All joints in the plasterboard, and joints between it and the floors, walls and soffit, would be sealed with an airtight mastic seal and the whole would be surfaced with several coats of a hard gloss paint. The fixings for the plasterboard would be dissociated from the old surface. The partition would now tend to reflect sound and reduce transmission, but by turning the sound back into the room the sound level would be considerably raised and transmission at the new high level is likely to be as objectionable as before. It is therefore doubly necessary to reduce the sound level itself in these outer offices.

Carpets and soft furnishings can go some way towards this, but extensive absorption treatment is probably necessary. This can be applied in several ways to both ceiling and walls. For the ceilings a variety of tile systems is available. Basically these consist of boards of insulating fibreboard, glass or mineral wool, or foamed polystyrene or perforated metal, plastics or plasterboard trays carrying glass or mineral wool infills. The results obtained will depend on the materials used.

Similar systems can be utilised for wall treatment and often it is necessary to attack the problem in stages rather than indulge in expensive treatment which may not all be necessary.

The director's office itself now has a new set of problems. If the treatments to the other offices have been successful, then the noise nuisances will have been removed, or at least reduced to acceptable proportions, but it may now be possible to hear, in the outer offices, conversations which take place in the director's room.

The remedial measure is again a containing treatment, the partition sides in the director's office being treated in the same way as were those in the other rooms. In normal circumstances there should be no necessity for the incorporation of sound absorption.

It will have been noticed that all the materials mentioned in this rather complex problem provide thermal as well as sound insulation and this would also be so if the use of these rooms had been anticipated at the design stage. Had the partitions been constructed in the first place from compressed straw slab cores faced with plasterboard and acoustic ceilings fitted in the outer offices, it is unlikely that the director would have been worried by noise.

In the reference to windows it has been implied that sound transmission is not greatly impeded by them. This is true of all single glazing and of double glazing where the cavity between the sheets is less than 4 inches wide. With a 2-inch cavity reasonable sound insulation is obtained provided that the sound levels involved are not very high, but really good results are not likely to be achieved unless a 4-inch or wider cavity is employed and the reveals within the cavity are treated with a sound absorbent.

Double glazing in these forms is, of course, a dual purpose material in the terms of this chapter.

Heating

In the heating field there are again a number of problems where dual purpose insulation materials have applications.

Where quick response heating systems are employed it is desirable, if not essential, that the inner linings of rooms so heated should be of low thermal capacity, thus ensuring that the benefit of the quick response heating is not lost as far as the occupants are concerned while the structure heats up. A lining of thermal insulation of fibreboard, glass or mineral wool or foamed polystyrene will overcome this difficulty while at the same time providing considerable sound absorption. Some of the available materials will require protection from mechanical damage, but this can be provided by the use of perforated sheets of materials which, although basically of high thermal capacity, can be used at a thickness which does not materially affect the results obtained.

Two particular examples of the use of dual insulation materials lie in the field of ducted warm air. In large industrial systems where high velocities of air are involved, the internal lining of the duct with a suitably protected insulation material has the dual function of reducing the level of noise and providing the necessary thermal insulation.

Similarly, in smaller domestic systems where a noise problem also exists, particularly with return air ducts, a new glass fibre duct has been developed to replace the metal ducts traditionally used and to provide both sound absorption and thermal insulation of a high order. In this type of system, employing in the main small ducts which cannot easily be insulated internally, the new technique offers by far the simplest way of overcoming the double problem.

Transport

In the transport industry also there are many problems which the dual purpose material can help to solve.

In a modern ship, which by its nature may be exposed to the full range of climatic conditions, the accommodation for passengers and crew alike must be of a reasonable standard of comfort. Where the main construction of cabins, etc., is of high thermal conductivity metal, one of the basic requirements is for thermal insulation; and it is also desirable that provision be made for quiet conditions.

Both of these requirements are met by the application of glass fibre or mineral wool boards to the bulkheads and deckheads. Some of these boards are available with special puncture-resistant finishes capable of withstanding the rigours of shipboard life, but in other cases highly decorative and tough finishes have been devised using slotted and perforated metals and wood.

Similar treatments are also used in many areas of the ship where the risk of condensation brings its own need for thermal insulation and the noise reduction obtained is an added advantage.

In aircraft the same problems are encountered, but they are very much more severe. Constant exposures to temperatures below the comfort level and to high pitch sound from the engines would make the passenger aircraft unbearable but for the use of very lightweight glass fibre blankets fitted in the cavities between the outer skin and the inside trimming fabric. The bulkheads are often similarly treated.

The same principles are also applicable to the modern motor car, where the constant thinning down of the metal body and the development of high performance engines have produced a wide range of thermal and sound problems. For many of these, flexible materials such as wool felt, glass and mineral wool blankets can provide the dual answer and in a few of the latest models a glass fibre one-piece moulded headliner complete with finished fabric surface is doing both jobs and also providing an attractive decorative appearance.

In the cab of a recently designed heavy-duty diesel-powered commercial vehicle also the same type of headliner is in use.

Similar techniques are also employed to improve conditions in the diesel locomotives now in service on our railways.

CONCLUSION

In this chapter the scope has been limited to those applications where dual purpose materials can be used. In the various industries there are, of course, many cases in which either sound or thermal insulation is required separately. It would be wrong to say that treatment for thermal insulation with one of these dual purpose materials automatically provides a bonus in the way of sound insulation. The reverse is true, however, and it is for this reason that sound insulation has been given prominence here, in a book which has as its title "Thermal Insulation."

8

The Design of Walls for Intermittent Furnaces Using High-temperature Insulation

W. H. HOLMES
The British Ceramic Research Association (Great Britain)

Although high-temperature insulating bricks have been available for thirty years or more they are still not as widely used as one might expect. In some cases, for example, where chemical or physical attack such as slagging, abrasion or erosion is likely they are, of course, unsuitable, and their use may also be questionable in atmospheres containing high concentrations of carbon monoxide, but some objections to their use are not well founded. For example, the statement that high-temperature insulating bricks cannot carry anything like the same load as a dense firebrick is true but scarcely relevant since the load imposed on a lightweight brick by overlying bricks of the same type is obviously much smaller than the load imposed on a firebrick by overlying firebricks. During the past ten or fifteen years, however, there has been a definite trend towards the use of high-temperature insulation. This has been due, first, to an improvement of these products in the direction of greater purity, refractoriness and mechanical strength together with a reduction in thermal conductivity, bulk density and heat storage capacity, and second, to the fact that furnace designers are learning how to make the most effective use of these materials.

There is no doubt that high-temperature insulation finds its most obvious application in furnaces that are operated intermittently since here, with its low thermal conductivity and low heat-storage capacity, it scores heavily over denser refractories. Such furnaces are constructed with composite walls and usually consist of either two or three different insulating materials. The high-temperature (but higher-thermal-conductivity) insulation forms the inner furnace lining, and this is backed by successively lower grades of material of lower thermal conductivity and therefore of increasing insulating qualities. These outer layers may be in the form of bricks, slabs or loose fill

WALLS FOR INTERMITTENT FURNACES

and their purpose is, of course, to reduce heat losses due to conduction. It is essential, however, not to use too great a thickness of this insulation as this can easily raise the temperature of the material next to the inner lining above its safe working level. Simple calculations, based on steady-state conditions, ensure that mistakes of this type are avoided, but they often lead to the use of furnaces that are unnecessarily bulky and that only cool in a reasonable time when air is drawn or blown through the interior. In practice the time–temperature schedules for pottery kilns are so short that steady-state conditions are never attained, and the introduction about 10 years ago of the "top hat" kiln, which had to be lifted and lowered, en-

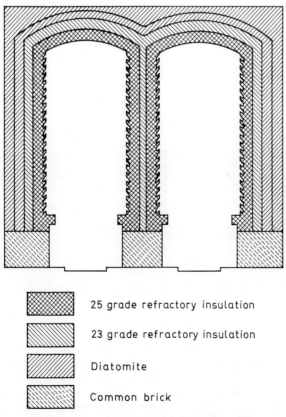

25 grade refractory insulation

23 grade refractory insulation

Diatomite

Common brick

Fig. 8.1. Twin-chamber electric intermittent kiln.

couraged more precise design which took this into account and therefore helped to eliminate all unnecessary weight.

Up to that time, and in some cases much more recently, intermittent kilns often had bulky multi-layer walls, and Fig. 8.1 shows a typical construction of this type. The inner lining consists of a $4\frac{1}{2}$-in thickness of a 25-grade brick, backed first by a 3-in layer of a lower grade high-temperature insulating brick, then by two 3-in layers of diatomaceous brick, and finally cased in sheet steel. This type of kiln was designed to fire bone china to a maximum temperature of 1250°C with a time cycle of 48 h.

Grooves to support the elements are cut from standard bricks and the ease with which any special shapes can be cut with simple tools is a most useful property of high-temperature insulating bricks.

Figure 8.2 is a photograph of a kiln constructed in this way. One unit consists of two chambers on a common electrical supply and these are heated alternately. Thus as one is being heated the other is

Fig. 8.2. Twin-chamber electric intermittent kiln.

cooling and *vice versa*. Previously loaded trucks can be pushed into or withdrawn from these chambers without having to wait for either the trucks or chambers to cool completely.

In 1955, however, a kiln of the top-hat type was introduced to fire bone china and Fig. 8.3 shows a more recent kiln of this type designed to fire porcelain electrical insulators. In the liftable cover or top-hat kiln the main heating elements are fitted in the walls and roof and this movable cover is lowered over a fixed base on which the goods are placed. The base can be unloaded and loaded whilst the cover is being used for firing and cooling on another base or whilst the cover is raised to permit free access as in Fig. 8.3.

With top-hat kilns it is obviously desirable to reduce the wall thickness to a minimum in order to make the movable covers as light as possible.

When the performance of one of these top-hat kilns with walls $7\frac{1}{2}$ in thick was compared with that of a truck-type kiln with $13\frac{1}{2}$-in walls it was found rather surprisingly that there was little difference in thermal

Fig. 8.3. Top-hat kiln for porcelain electrical insulators.

efficiency. This aroused considerable interest since, apart from the fact that top-hat kilns must be as light as possible, it is also desirable to reduce the size and weight of any furnace in order to reduce building costs and weight on foundations, and also to facilitate the transport of prefabricated furnaces. It also raised such questions as "How can $7\frac{1}{2}$-in walls be as effective in reducing heat losses as $13\frac{1}{2}$ in walls?"

The answer to this particular question did not prove to be simple such as finding that the thinner wall consisted of better insulating materials. There was no doubt that the thicker wall had an appreciably greater thermal resistance, and a detailed study of heat flow through walls of this type was made in order to answer such queries, and to provide a sound basis for the design of composite walls for intermittent furnaces.

After various possibilities such as thermal and electrical models had been considered the problem was tackled by making calculations based on the numerical methods developed and expounded by Dusinberre[1,2]. In these methods the furnace structure is represented by a finite network of points, and temperatures are calculated only for these preselected points. The only facts necessary are: the thermal properties of the various types of bricks and other materials used, the time–temperature schedule and the ambient temperature. Also, whereas the methods of calculus require a high degree of mathematical training, there is nothing more complicated than addition, multiplication and division in the actual worksheets of the numerical method.

This method, which is described briefly in an appendix to this chapter, was used to calculate the temperature distribution in a four-layer wall of an electric intermittent kiln firing bone china to 1240°C, and the results were verified by inserting thermocouples at the brick interfaces during the building of such a kiln and obtaining a complete record of the temperatures when the kiln settled down in regular cyclic operation. The wall construction was the same as that of the truck-type intermittent kiln shown in Fig. 8.1 but, because the element grooves were $1\frac{1}{2}$ in deep, an effective thickness of 3 in instead of $4\frac{1}{2}$ in was taken in the calculations for the inner lining. The calculated and experimental results have been compared in Fig. 8.4, from which it can be seen that there was a very satisfactory agreement between the calculated results and those obtained by direct measurement. The other important point to note is that the multilayer-wall system led to interface temperatures that were hundreds of degrees within the safe working limit of the backing insulation. The first interface, for example, reached little over 900°C whereas the backing insulation is capable of withstanding 2300°F or 1260°C and the third interface was 370°C whereas the diatomaceous bricks can be used up to 870°C.

The good agreement between the calculated and the experimental results engendered confidence not only in the method of calculation but also in the published thermal properties of the insulating bricks, with the result that other possible wall constructions were examined theoretically using the same procedures.

The first of these was similar to one that had already been used successfully in top-hat kilns. This consisted of $4\frac{1}{2}$ in of a 25-grade high-temperature insulating brick backed by 3 in of calcium silicate slab insulation. The calculated temperature distribution is shown in Fig. 8.5. The principal result was that the maximum temperature at the interface between the 25-grade brick and the slab insulation was 970°C compared with the manufacturer's stated safe limit of 1010°C for the slab insulation.

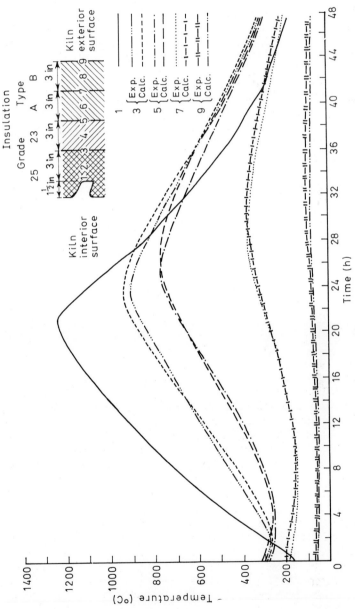

Fig. 8.4. Four-layer wall. Calculated vs. experimental.

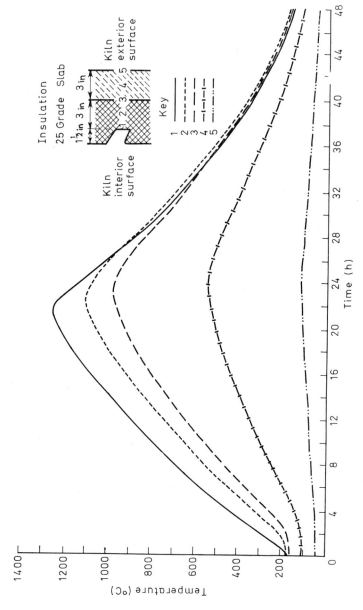

Fig. 8.5. Two-layer wall calculation. 4½ inches of 25 grade and 3 inches of slab.

This work indicated, therefore, that such a wall is suitable for a maximum temperature of 1240°C and a 48-h cycle. In fact top-hat kilns with similar walls consisting of $4\frac{1}{2}$ in of a 28-grade brick backed by 3 in of calcium silicate slab insulation had been built prior to this work and used successfully for the firing of bone china to similar temperatures with a 48-h cycle.

A further calculation was made for a $7\frac{1}{2}$-in wall, again using $4\frac{1}{2}$ in of a 25-grade brick but backed this time by 3 in of a diatomaceous brick. The results of this work are shown in Fig. 8.6. Again the chief interest lay in the maximum temperature at the interface between the 25-grade brick and the diatomite backing. This was almost 900°C compared with the manufacturer's stated safe limit of 870°C for the diatomaceous brick which indicates that a wall of this composition would be unsuitable for temperatures of 1240°C and 48-h cycles. The results of the calculations were also used as a basis for calculating the heat losses due to conduction through and heat storage in the walls. This work was being done in order for a comparison to be made between a four-layer and a two-layer wall.

The manner in which heat losses occur in an intermittent furnace is not always as simple as it might appear at first sight. During the heating period heat is flowing mainly in one direction through the furnace structure, that is, into the interior surface and through the wall towards the exterior surface. Some of this heat is lost from the exterior surface by radiation and convection and the remainder is stored in the wall. If the furnace were sealed during the cooling period all the heat stored in the structure would have to be dissipated from the exterior surface by radiation and convection. In practice, of course, this would usually result in an unacceptably long cooling time and air is therefore often either blown or drawn through the furnace in order to cool the goods and the interior surfaces. Since the back-up insulation has a lower thermal conductivity and a lower heat capacity than those of the hot-face insulation most of the heat stored in the furnace structure at the end of the heating period is in the refractories at, or near to, the interior surface. This means that during cooling more heat can be lost from the interior than from the exterior. For the actual kiln that was used to verify the method of calculating the temperature distribution the situation was complicated slightly by the fact that a truck could be withdrawn and replaced by another at kiln temperatures over 200°C. Thus the truck was in the kiln chamber for 2 h 40 min before the kiln was switched on, and during this period heat stored in the brickwork was being utilised to heat the truck and its goods. In this case, therefore, the

102 THERMAL INSULATION

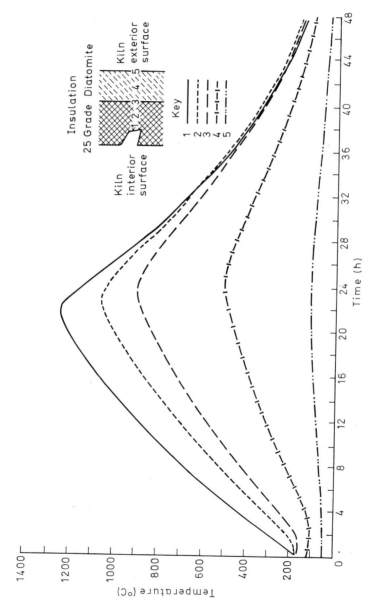

Fig. 8.6. Two-layer wall calculation. 4½ inches of 25 grade and 3 inches of diatomaceous brick.

heating period has been taken as from the time when the truck entered the kiln to the time when the kiln was switched off, the cooling period being taken as the remainder of the cycle. The kiln was operated on a 48-h cycle and the time that it was switched on has been taken as zero since this was the start of the programme-controlled heating period. The kiln was switched off after 22 h and the truck withdrawn and replaced by another after 45 h 20 min. Thus the heating period was 24 h 40 min and the cooling period 23 h 20 min.

Heat losses were assessed by calculating the radiation and convection losses from one square foot of the exterior surface during heating and cooling and by also calculating the heat stored in the corresponding volume of the wall at the end of the heating and at the end of the cooling period. Table 1, which shows how the heat losses through one of the two-layer walls were derived, should help to make this clear.

TABLE 1

HEAT LOSSES THROUGH TWO-LAYER WALL
$4\frac{1}{2}$-IN, 25-GRADE, 3 IN SLAB

Method of heat lost	Btu/ft^2
Heat gained by wall during heating period	8,215
Loss from exterior surface during heating period	3,530
Heat flow into wall during heating period	11,745
Heat lost by wall during cooling period	8,215
Loss from exterior surface during cooling period	4,035
Loss from interior surface during cooling period	4,180

The first item, the heat gained by the wall during the heating period, is the difference in the heat contents of the wall during the period involved. The sum of this and the loss from the exterior surface during the same period gives the total heat flow into the wall during this period. The heat lost by the wall during the cooling period is, of course, the same as that gained during the heating period since the kiln is in regular cyclic operation and temperature conditions are repeated every 48 h. Part of this is lost from the exterior and part from the interior surface. The loss from the exterior surface is calculated directly and the loss from the interior surface obtained by difference.

The heat losses for the four-layer wall were assessed in similar manner and the results have been compared with those for the two-layer wall in Table 2.

TABLE 2

COMPARISON OF HEAT LOSSES THROUGH TWO- AND FOUR-LAYER WALLS

	Two-layer wall (Btu/ft^2)	Four-layer wall (Btu/ft^2)
Flow into wall during heating period	11,745	12,515
Loss from exterior surface during heating period	3,530	1,600
Loss from exterior surface during cooling period	4,035	2,815
Loss from interior surface during cooling period	4,180	8,100

This comparison shows that although roughly the same amount of heat flows into each of the walls during heating the manner in which it is lost is very different. Thus over twice as much heat passes through the thinner, two-layer wall during heating to be lost from the exterior surface by radiation and convection, and considerably more again passes through during cooling to be lost in the same way. But the thicker, four-layer wall loses most of its heat from the interior surface during cooling. These figures can be summed up quite simply by saying that although roughly the same amount of heat enters both types of wall to be lost eventually one way or another, the thinner, two-layer wall allows about two-thirds to pass through to be lost from the exterior surface, whilst the thicker, four-layer wall only allows about one-third to pass through, the remaining two-thirds being lost from the interior surface to air which is invariably drawn through such kilns to assist cooling.

Materials such as ceramics are very subject to thermal failure if cooling is too rapid over certain temperature ranges and there may be a considerable risk in drawing cold air through the interior of kilns, but it is essential in the interests of productivity to achieve a rapid kiln cycle. For some products, therefore, one designer has deliberately reduced wall thicknesses below the values required for optimum thermal efficiency in order that cooling should proceed at a reasonably fast rate without using air to cool the interior.

To sum up, it has proved possible by means of a numerical method to calculate the interface temperatures in composite walls and to assess the heat losses for any time/temperature cycle. At the British Ceramic Research Association this method has been used to design a 2000°C furnace with four-layer walls and there seems little doubt that when once the basic principles have been firmly grasped a working scheme can be readily devised for any situation.

REFERENCES

1. G. M. Dusinberre, *Trans. A.S.M.E.*, **67** (1945) 703.
2. G. M. Dusinberre, *Numerical Analysis of Heat Flow*, McGraw-Hill, New York, 1949.

APPENDIX

METHOD OF CALCULATING TEMPERATURE DISTRIBUTION

The method used was based on the principles given by Dusinberre in "Numerical Analysis of Heat Flow" (McGraw-Hill, 1949). The following is an attempt to present a concise account of these principles and to show how they have been applied in this particular case.

Consider a part of a wall far enough away from the edges for the heat flow to be one-dimensional. In Fig. 1A this portion of the wall has been divided into cubical blocks of side length Δx with the points 1, 2, 3, etc., at the centres of these cubes. The size of this increment

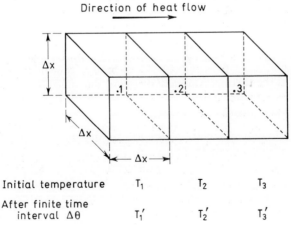

Fig. 1A. Division of portion of a wall into cubical blocks.

Δx, into units of which the wall thickness is divided, is chosen to suit the requirements of the particular problem. The temperatures at the points 1, 2, 3, etc., are assumed to be representative of their respective cubes. The temperatures at a certain time, in this case taken as the time when the furnace elements were switched on, are guessed initially. All previous knowledge or any theoretical considerations can be incorporated in this guess and the better the guess the less the subsequent calculations. These initial temperatures at points 1, 2, 3, etc., are denoted as T_1, T_2, T_3, etc., and the basis of the method is to calculate the temperature at these points after a finite time interval $\Delta\theta$. The temperatures at the end of such a time interval are denoted as T_1^1, T_2^1, T_3^1, etc.

It is assumed that the time interval $\Delta\theta$ can be chosen sufficiently small that the temperature at point 2 is not affected by any change at the left of cube 1 and at the right of cube 3. Thus only T_1, T_2 and T_3 need be considered in calculating T_2^1. It is also assumed that the time interval is small enough to allow the initial temperature gradients to be used for the whole of the interval and for the change of heat storage in a cube to be calculated from the change of temperature of the point at its centre.

Making use of these assumptions we can write a heat balance for an interior point such as 2 as follows:

The heat flow from cube 1 to cube 2 is given by:

$$\left(\frac{(\Delta x)^2 (T_1 - T_2) k . \Delta\theta}{\Delta x}\right) - \left(\frac{(\Delta x)^2 (T_2 - T_3) k \Delta\theta}{\Delta x}\right)$$
$$= c\rho(\Delta x)^3 (T_2^1 - T_2) \qquad (1)$$

or $\quad (T_1 - T_2) - (T_2 - T_3) = \dfrac{(\Delta x)^2}{k\Delta\theta}(T_2^1 - T_2)$

where k is the thermal conductivity
$\quad\quad c$ is the specific heat
and ρ is the density.

This equation is greatly simplified by defining a dimensionless product, called a modulus, as

$$M = \left(\frac{c\rho(\Delta x)^2}{k\Delta\theta}\right)$$

Substituting this in the above equation and rearranging we get the simple form

$$T_2^1 = \frac{T_1 + (M - 2) T_2 + T_3}{M} \qquad (2)$$

and when $M = 2$ we obtain the very simple form

$$T_2^1 = \frac{T_1 + T_3}{2} \qquad (3)$$

which is the numerical equivalent of Schmidt's graphical method.

From the defining equation for M it will be seen that there are three variables involved, Δx, $\Delta\theta$ and M. In practice Δx is usually fixed, therefore the choice of θ determines M or *vice versa*. Some experience is required to make a good choice of value for $\Delta\theta$ or M. Large values of M lead to unbearably long calculations but values can be too small and can lead to divergent oscillations. It is often convenient to fix Δx and $\Delta\theta$ and, in this case, eqn. (2) is usually expressed in the form

$$T_2^1 = F_{12}T_1 + F_{22}T_2 + F_{32}T_3 \tag{4}$$

where

$$F_{12} \text{ and } F_{32} = \frac{1}{M} \text{ and } F_{22} = \frac{M-2}{M} \text{ and } \Sigma F = 1$$

so that this method of calculating T_2^1 can be regarded as a weighting procedure in which the three initial temperatures T_1, T_2 and T_3 are each multiplied by a fraction and the products added. The first subscript of F refers to the influencing and the second to the influenced temperature.

Calculations for composite walls can be made by this method since the F values can be calculated for a point at the interface between two different materials without explicitly using the two corresponding M values. The heat balance for the point in question is written as in eqn. (1) using the two different values of k on the left-hand side and an average for the two different values of $c\rho$ on the right-hand side.

For most insulating materials the thermal conductivity and the specific heat vary independently with temperature to an appreciable extent and to allow for this the F values were recalculated every 6 h in a 48-h cycle.

An idea of the work involved can be gauged from the worksheet for the first 6 h of a two-layer wall calculation shown in Table 1A.

This calculation is for a wall with a 3-in thickness of high-temperature insulating brick backed by a 3-in thickness of slab insulation. Δx has been taken as $1\frac{1}{2}$ in so that the 6-in thick wall is divided into three cubical blocks in the interior and a half cube at the internal and external surfaces. $\Delta\theta$ has been taken as 20 min so that temperatures are calculated at 20-min intervals. The values of T_1, which represent the temperatures of the hot face of the wall, are taken from the known time–temperature schedule for the furnace. The values of T_2 and T_4, which are the temperatures at the centres of the two materials, are both calculated exactly as described above. The fraction 0·324 is the F value of both F_{12} and F_{32}, which is the reciprocal of the modulus calculated for the high-temperature insulating brick, and F_{34} and F_{54} also having the same value 0·252, which is the reciprocal of the modulus calculated for the slab insulation. T_3 is the temperature that is usually of most interest since it is the interface temperature between the two materials. This is calculated as described above by using whichever of the two values of k is appropriate and the average value for c. Thus F_{23} and F_{43} do not have the same values. The calculation of T_5, which is at the external surface of the wall,

TABLE 1A

CALCULATIONS FOR TWO-LAYER WALL. WORKSHEET FOR FIRST 6 HOURS

Time (h min)	T_1 (°C)	T_2 0·324 T_1 +0·352 T_2 +0·324 T_3 (°C)	T_3 0·510 T_2 +0·370 T_3 +0·120 T_4 (°C)	T_4 0·252 T_3 +0·496 T_4 +0·252 T_5 (°C)	T_5 0·198 T_4 +0·802 T_6 (°C)	T_6 (°C)
0·00	160	180	200	150	50	30
0·20	180	180	184	137	51	30
0·40	195	181	176	127	49	30
1·00	215	184	173	120	48	30
1·20	240	190	172	115	47	30
1·40	265	200	174	112	46	30
2·00	290	213	180	111	46	30
2·20	315	227	189	112	46	30
2·40	345	243	199	115	47	30
3·00	370	262	211	119	48	30
3·20	390	280	226	124	49	30
3·40	410	298	241	131	50	30
4·00	435	316	257	138	51	30
4·20	460	335	273	146	53	30
4·40	480	355	289	155	55	30
5·00	505	374	307	164	57	30
5·20	520	395	324	173	58	30
5·40	540	412	342	182	60	30
6·00	565	431	359	192	62	30

requires rather different treatment. In this case an approximation involving the use of a steady-state equation has been used. This ignores the heat capacity of the half-cube at the surface, but this was shown to be justified by the fact that the temperatures near to the surface change very slowly. It must be remembered that in this current and not past temperatures are used in calculating T_5[1].

Summing up, the values for T_1 are obtained from the time-temperature schedule. The ambient temperature, T_6, can be guessed fairly accurately; the other temperatures at zero or switch-on time are guessed and then all the other figures are computed using these F values. For example, to obtain T_4 after 20 minutes, T_3, which is 200, is multiplied by 0·252; T_4, which is 150, is multiplied by 0·496 and T_5, which is 50, is multiplied by 0·252. The sum of these products is 137, which is T_4^1 or the temperature of T_4 after an interval of 20 minutes. T_2 and T_3 are computed in the same way, but T_5 is computed from current values of T_4 and T_6; thus T_4^1, which is 137, is multiplied by 0·198 and T_6^1, which is 30, is multiplied by 0·802 and the sum of these products is 51. This simple arithmetic is continued until the worksheet has been completed and then work continues with the second worksheet which has different F values to allow for changes of thermal conductivity and specific heat with temperature. When a 48-h cycle has been completed the initial guesses are compared with the results obtained by this first approximation. The whole cycle could now be recalculated to obtain greater accuracy, but experience leads to such good initial guesses that this is seldom necessary.

9

Selection and Comparison of Materials Using Figures of Merit

P. H. H. BISHOP

C.P.M. Department, Royal Aircraft Establishment, Farnborough
(Great Britain)

INTRODUCTION

Selection of materials depends on many factors, but some aspects of selection can be made easier by the use of "figures of merit". These are simple functions of material properties, and although their use has been confined mainly to aircraft design, they have many wider applications when structural and thermal requirements occur simultaneously. Frequently they provide direct comparison of cost, weight and heat losses for the same project constructed from different materials. They may have many applications to instrument design when there is an interplay of weight, thermal and electrical requirements. For the materials specialist they indicate which properties should be improved.

This chapter shows how figures of merit are derived, and gives examples. The metric system of units is employed, to illustrate more clearly the quantitative significance of figures of merit. For comparative purposes, the system of units employed does not matter.

In the design of every system there is of course an overriding aim, *e.g.*, minimum weight or cost, maximum efficiency or some combination of these. Every system can be thought of as developing and selecting the best materials, and shaping and putting them together in the best way.

In the design of most systems, the loads on local areas of material are calculated, then the material is chosen (probably from a class already decided) and designed to have say minimum local volume, weight or heat loss. The simplest system is a bar to resist tension load. For a bar of minimum weight, one material is best; for minimum cost, another; and for minimum axial heat flow, yet another. The strongest material will give minimum volume but not necessarily minimum weight. The most efficient material is determined by a function of several properties, called a "figure of merit". These functions are frequently simple ratios, *e.g.*, strength/density, strength/thermal conductivity. They are also known as merit indices.

LIST OF SYMBOLS AND UNITS

E	Young's modulus (kg/cm^2)
G	Shear modulus (kg/cm^2)
f	Strength defined in any appropriate way; ultimate tensile strength in this note (kg/cm^2)
f_b	Modulus of rupture (kg/cm^2)
f_s	Shear strength (kg/cm^2)
d	Density (g/cm^3)
k_1	Thermal conductivity parallel to axis (cal/s cm deg C)
k_2	Thermal conductivity through thickness (cal/s cm deg C)
ρ_1	Electrical resistivity parallel to axis ($\mu\Omega$ cm)
ρ_2	Electrical resistivity through thickness ($\mu\Omega$ cm)
c	Specific heat (cal/g deg C)
α	Linear thermal expansion (deg C^{-1})
t	Time (s)
θ	Temperature (°C)
θ_1, θ_2	Temperature of opposite ends of rod or faces of panel (°C)
M	Bending moment (kg cm)
T	Tensile load (kg)
W	Weight of shaped material (g)
C	Cost of shaped material (any arbitrary cost units)
Q	Heat flow (cal/s)
I	Current (amps) or moment of inertia (kg cm^2) according to context
A	Area of cross-section (cm^2)
l	Length of panel or rod (cm)
h	Breadth of panel (cm)
b	Thickness of panel (cm)
r	Radius of rod (cm)
C_o	Approximate cost of material (any arbitrary cost unit per g)

Suffixes 1 and 2 are employed to avoid ambiguity where necessary:
 1 — Axial
 2 — Through thickness

P	General physical property
M	General mechanical property
n	Exponent

SIMPLE FIGURES OF MERIT

Materials must usually resist load without rupture or excessive deflection. Area of cross-section is calculated from dimensions, loads, mechanical properties and specified limitations on behaviour. This area automatically defines the interesting general features of the shaped material, *e.g.*, its weight or cost.

Many figures of merit are derived by such procedures as (a) calculating area of cross-section to meet a mechanical requirement and (b) combining this area with physical properties to calculate a feature of interest. Assume that a lightweight tension bar has to be designed. Then:

(a) $A = T/f$
(b) $W = dlA = (lT)(d/f)$.

The group (lT) pertains to scale, the group (d/f) pertains to material properties and is the figure of merit. Suppose now that the main interest is heat flow along the bar, then:

(a) $A = T/f$
(b) $Q = Ak(\theta_1 - \theta_2)/l = T(\theta_1 - \theta_2)/l . (k/f)$.

The figure of merit is now k/f.

The importance of figures of merit lies in their capacity to compare directly how much a project weighs, costs, . . . loses heat, when constructed from different materials. They can be used thus because scaling factors are independent of material. Thus, a figure of merit consolidates several properties into a single number which shows directly what the material can do.

GENERAL FIGURES OF MERIT

Ordinary properties are usually defined in terms of a material phenomenon referred to unit cross-sectional area. Strength is "failing load per unit area of cross-section".

Figures of merit may be thought of as generalised properties in which one phenomenon is referred directly to another; area is not mentioned explicitly. Thus there could be material "properties" such as "tensile strength per unit weight" and "axial stiffness per unit cost".

Derivation of one figure of merit

General figures of merit may be derived from classical formulae. Thus for "Axial heat flow per unit bending stiffness for round bar" the *classical formulae* are:

$$M = EI/R, \quad I = \tfrac{1}{4}\pi r^4, \quad Q = Ak(\theta_1 - \theta_2)/l, \quad A = \pi r^2.$$

Then:

(a) *Calculation of area*

$$I = MR/E, \quad r^2 = (4I/\pi)^{\frac{1}{2}},$$
$$A = \pi r^2 = \pi(4I/\pi)^{\frac{1}{2}} = \pi(4MR/E\pi)^{\frac{1}{2}}.$$

(b) *Calculation of feature from area*

$$Q = \pi(4MR/E\pi)^{\frac{1}{2}} \cdot k(\theta_1 - \theta_2)/l$$
$$= (\theta_1 - \theta_2)/l \cdot (4\pi MR)^{\frac{1}{2}} \cdot (k/E^{\frac{1}{2}}).$$

Thus the figure of merit is $k/E^{\frac{1}{2}}$.

Figures of merit associated with axial phenomena have the form $P^{\pm 1}/M^n$. They are summarised in Table 1, in which mechanical requirements are defined across the top, and physical features are shown in the left-hand column.

Panels and thin-walled tubes have in general equivalent figures of merit.

Numerical examples

	d/E	$d/E^{\frac{1}{3}}$	d/f	$d/f^{\frac{1}{2}}$
Beryllium	6.5×10^{-7}	1.3×10^{-2}	5.2×10^{-4}	3.1×10^{-2}
Sintered alumina	1.0×10^{-6}	2.4×10^{-2}	1.4×10^{-3}	6.4×10^{-2}
Glass fibre laminate	9.3×10^{-6}	3.2×10^{-2}	4.9×10^{-4}	2.8×10^{-2}
	k_1/E	$k_1/E^{\frac{1}{3}}$	k_1/f	$k_1/f^{\frac{1}{2}}$
Beryllium	1.3×10^{-7}	2.5×10^{-3}	1.0×10^{-4}	6.1×10^{-3}
Sintered alumina	3.3×10^{-8}	7.8×10^{-4}	4.6×10^{-5}	2.0×10^{-3}
Glass fibre laminate	4.7×10^{-9}	1.6×10^{-5}	2.5×10^{-7}	1.4×10^{-5}

(For simplicity materials are assumed to be isotropic, in particular $k_1 = k_2$.)

A glass fibre laminate beam is $9.3 \times 10^{-6}/6.5 \times 10^{-7} = 14.3$ times heavier than a beryllium beam of equivalent axial stiffness.

TABLE 1
FIGURES OF MERIT FOR AXIAL PROPERTIES

$$\text{Mechanical requirement} \begin{pmatrix} \mathbf{S} = \begin{matrix} Panels, \\ sheet, \\ thin\ walled\ tube \end{matrix} & \mathbf{B} = \begin{matrix} Round\ bar, \\ rod\ or\ wire \end{matrix} \end{pmatrix}$$

	Stiffness								Strength					
	Axial		Buckling Bending		Shear		Torsion		Axial		Bending		Shear	Torsion
Feature of interest	S	B	S	B	S	B	S	B	S	B	S	B	S	B	S	B
	$1/E$	$1/E$	$1/E^{\frac{1}{3}}$	$1/E^{\frac{1}{2}}$	$1/G$	$1/G$		$1/G^{\frac{1}{2}}$	$1/f$	$1/f$	$1/f_b^{\frac{1}{2}}$	$1/f_b^{\frac{2}{3}}$	$1/f_s$	$1/f_s$	—	$1/f_s^{\frac{2}{3}}$

Feature of interest	
Volume	1
Weight	d
Cost	dC_0
Heat flow	k_1
Current	$1/\rho_1$
Heat capacity	c

Examples: Figure of merit = factor in above horizontal row × factor in left hand vertical column

(a) Cost per unit bending stiffness for round bar = $\dfrac{dC_0}{E^{\frac{1}{2}}}$

(b) Current flow per unit axial strength for sheet = $\dfrac{1}{\rho_1 f}$

Transverse properties

Sometimes high transverse flows of heat or electric current are required. The figure of merit for transverse heat flow through a panel under axial tension is derived thus:

(a) $T = hbf$; $b = T/hf$.
(b) $Q = lhk_2 (\theta_1 - \theta_2)/b = lhk_2 (\theta_1 - \theta_2) hf/T$
    $= lh^2\{(\theta_1 - \theta_2)/T\} (k_2 f)$.

Figure of merit is then $k_2 f$. These figures of merit have the form $P^{\pm 1} M^n$ and are summarised in Table 2. They are applicable to panels or thin-walled tubes.

TABLE 2
FIGURES OF MERIT IN WHICH TRANSVERSE PHYSICAL BEHAVIOUR IS OF PREDOMINANT INTEREST

| | Mechanical requirement | | | | | |
| | Stiffness | | | Strength | | |
	Axial	Bending	Shear	Axial	Bending	Shear
Transverse heat flow	$k_2 E$	$k_2 E^{\frac{1}{3}}$	$k_2 G$	$k_2 f$	$k_2 f_b^{\frac{1}{2}}$	$k_2 f_s$
Transverse current	$E/\rho 2$	$E^{\frac{1}{3}}/\rho 2$	$G/\rho 2$	$f/\rho 2$	$f_b^{\frac{1}{2}}/\rho 2$	$f_s/\rho 2$

Example

	$k_2 E$	$k_2 E^{\frac{1}{3}}$	$k_2 f$
Beryllium	$1 \cdot 0 \times 10^7$	$5 \cdot 1$	$1 \cdot 3 \times 10^3$
Sintered alumina	$4 \cdot 4 \times 10^5$	$1 \cdot 8$	$3 \cdot 1 \times 10^2$
Glass fibre laminate	$1 \cdot 9 \times 10^2$	$5 \cdot 6 \times 10^{-2}$	$3 \cdot 7 \times 10^1$

FIGURES OF MERIT FOR CURRENT AND HEAT FLOW

Some useful figures of merit may be calculated independently of mechanical properties. Areas of cross-sections are calculated and then other features are worked out.

Unit axial electrical conductance

(l fixed) Conductance $= A/\rho_1 l = 1$; $A = \rho_1 \quad l$; vol $= Al$
Volume per unit axial electrical conductance $= (l^2)\rho_1$
Weight per unit axial electrical conductance $= (l^2)d\rho_1$
Cost   per unit axial electrical conductance $= (l^2)C_o d\rho_1$

Axial heat flow per unit axial electrical conductance
$$= \text{(Scaling factor)} \times (k_1 \rho_1)$$
Heat capacity per unit axial electrical conductance
$$= \text{(Scaling factor)} \times (cd\rho_1)$$

Unit axial thermal conductance
By analogy
Volume per unit axial thermal conductance = (scaling factor) × l/k_1
Weight per unit axial thermal conductance = (scaling factor) × d/k_1
Cost   per unit axial thermal conductance = (scaling factor)
$$\times C_o d/k_1$$
Heat capacity per unit axial thermal conductance = (scaling factor)
$$\times cd/k_1$$

Unit axial electrical resistance
("A" fixed) Resistance = $\rho_1 l/A = 1$; $l = A/\rho_1$;   vol. = Al
Volume per unit axial electrical resistance = $(A^2) \times l/\rho_1$
Weight per unit axial electrical resistance = $(A^2) \times d/\rho_1$
Cost   per unit axial electrical resistance = $(A^2) \times c_o d/\rho_1$
Heat capacity per unit axial electrical resistance
$$= (A^2) \times cd/\rho_1$$
Axial heat flow per unit axial electrical resistance
$$= \text{(scaling factor)} \times k_1/\rho_1$$

Unit axial thermal resistance
Volume per unit axial thermal resistance = (scaling factor) × k_1
Weight per unit axial thermal resistance = (scaling factor) × dk_1
Cost   per unit axial thermal resistance = (scaling factor) × $C_o dk_1$
Heat capacity per unit axial thermal resistance
$$= \text{(scaling factor)} \times cdk_1$$

Numerical examples
(a) *Lightweight electrical resistances* (d/ρ_1)
    Titanium alloy 3×10^{-2}
    Stainless steel 7×10^{-2}

(b) *Lightweight electrical conductors* ($d\rho_1$)
    Beryllium   7·4
    Al alloy   10
    Copper   15

SELECTION AND COMPARISON OF MATERIALS

(c) *Lightweight thermal insulators* (dk_1)
 Micro glass fibre felt $4 \cdot 3 \times 10^{-6}$
 Balsa $4 \cdot 1 \times 10^{-5}$
 Glass fibre laminate $2 \cdot 2 \times 10^{-4}$

(d) *Lightweight thermal conductors* (d/k_1)
 Sintered beryllia $3 \cdot 2$
 Beryllium $5 \cdot 1$
 Copper $9 \cdot 8$
 Silver 10

(e) *Axial heat flow per unit axial conductance* $(k_1 \rho_1)$
 Iridium $0 \cdot 75$
 Cobalt $1 \cdot 0$
 Copper $1 \cdot 5$
 Silver $1 \cdot 6$
 Lead $1 \cdot 7$

Many experiments require delivery of power with minimum heat loss; from (e) above it is seen that cobalt-based power leads would conduct 37% less heat away than silver-based leads delivering the same current.

SUBSIDIARY FIGURES OF MERIT

Further figures of merit are presented without derivation.

Constrained panel with steady thermal gradient
Surface stress: $E\sigma$
Maximum temperature differential: $F/E\sigma$
Maximum heat flux (Biots modulus): $kf/E\sigma$

Constrained panel suddenly heated on one face
Maximum surface stress: $E\sigma\, k/dc$
Maximum temperature change $\times$ (time)$^{\frac{1}{2}}$: $cfd/E\sigma k$

Miscellaneous
Heat leak into interior: $(kdc)^{\frac{1}{2}}$
Thermal diffusivity: (k/dc)
Rate of temperature rise under unit potential: $l/\rho dc$
Cost per unit heat capacity: C_o/c

Minimum temperature lag

If the temperature on one face of a panel is changed or cycled, the temperature of the other face will lag by an amount proportional to $b^2/(k_2/dc)$. Figures of merit for minimum lag may be derived by eliminating b (e.g., for weight, $b = l/d$).

Transverse temperature lag per unit volume or thickness: dc/k_2
Transverse temperature lag per unit weight: c/dk_2
Transverse temperature lag per unit cost: $c/C_o^2 dk_2$
Transverse temperature lag per unit axial stiffness: $dc/E_1^2 k_2$
Transverse temperature lag per unit axial strength: $dc/f_1^2 k_2$
Transverse temperature lag per unit bending stiffness: $dc/E^{\frac{1}{3}} k_2$
Transverse temperature lag per unit bending strength: dc/fk_2
Transverse temperature lag per unit axial resistance $= \rho_1^2 dc/k_2$
Axial temperature lag per unit axial conductance $= dc/\rho_1^2 k_1$

Multicondition figures of merit

Mechanical and physical features may not be required at the same ambient condition. Figures of merit may be derived such as k_{100}/f_{500}, meaning: "Axial heat flow at 100°C per unit axial load at 500°C". This figure of merit would have application to design of supports in which the main mechanical load occurred at 500°C, the main heat loss at 100°C.

DISCUSSION

The main problem associated with figures of merit is the task of producing them. The author has written tentative computer programmes for converting miscellaneous materials information into figures of merit presented in tabular form. The computing costs are by no means insignificant and the task of gathering the materials information in the first place is of course formidable.

However, there may be sound economic arguments for facing up to this task, for figures of merit may point the way to useful savings in materials. By comparing materials in precisely the form the designer wants, the figures of merit may reveal unsuspected technical possibilities.

Index

Absorbed radiation, 66
Adsorbent, 25
"Afterthought" insulation, vii
Air, 45
Airborne sound, 85
Air cells, 38
Airspace resistance, 69
Aluminium alloy, 116
Aluminium panels, 51, 52
Ammonia, 43
Asbestos, 51

Balsa, 117
Beryllium, 113, 115–117
Blinds, 76
Brick: 25 grade, 96, 98
Building panels, 58
Burt Committee, 83
Butadiene, 42

Caltrop, 31
Cardboard, 85
Carbon dioxide, 45
Ceramics, 104
Chevron Hilton hotel, 49, 51
Chlorofluorohydrocarbon, 46, 48
Cobalt, 117
Composite walls, 94
Condensation, vii, 77
Contact resistance, 29, 36
Convection, 17
Copper, 116, 117
Cork, 38, 47
Cost, 112
Cryocable, 26
Cryogenic engineering, 15, 17
Curtains, 76

Dessicant, 78
Design, vii
Designer, 118
Diatomaceous brick, 96
Diatomite, 95, 101
Diisocyanatodiphenylmethane, 57
Distillation column, 6
Double glazing, 74, 75, 91
— —, "do-it-yourself", 80
Duplex dewar, 10
Dusinberre numerical methods, 98, 105–109

Equivalent temperature (ET), 81
Evacuated powder reflector, 17
Expanded ebonite, 7, 38, 40, 42, 47
— perlite, 5
— polystyrene, 38, 40, 41, 45, 47
— polyurethane, 44, 45, 48, 49, 56
— PVC, 7, 39, 47

Fibreboard, 47, 88
Fibres, 38
Figure of merit, 110
— — — ($k\rho$), 26
Firebrick, 94
Fish, 41
Flammability, ix
Floating floor, 86
Food, 40
Free electrons, 29
Fruit, 40, 41
Furnace walls, 94

Glass, 62, 65, 69
Glass fibre, 47, 92
Glass fibre duct, 92

Glass fibre laminate, 113, 115, 117
Glass surface temperatures, 71
Glass wool, 87, 88, 90
Glazing, 67
Global Radiation, 64

Hermetically sealed units, 77, 79
High temperature insulation, 94
Holiday chalet, 55

Ice-skating rinks, 43
Industrialized building, x
Infra red, 64
Injection technique, 57, 59
Insulated trailer, 60
Insulating blankets, 92
Insulating bricks, 94
Insulants—desirable characteristics, viii, 3, 4, 39
Iridium, 117

Jointless insulation, 48, 60

Lamination, 85
Lead, 117
Liquefaction, power expenditure, 5
—, costs, 16
Liquid refrigerants, vii, 1, 4, 16
Loudness, 84
Low-temperature insulation, 1

Material selection, 110
Mean free path, 18
Mean radiant temperature (MRT), 81
Meat, 40, 41
Methane, 41
Microcel, 13
Microglass fibre felt, 117
Mineral wool, 92
Mosaic tile cladding, 50
Multilayer insulants—superinsulations, 11, 15
Multilayer supports, 29
Multilayer walls, 96, 103, 104

Multiple glazing, 62
— —, economics, 80

Natural radiation, 63
Noise, 84
Noisy neighbours, 86

Optimum airspace width, 70
— packing density, 7
— thickness, x

Perlite, 17, 22
Permeability, 38
Phonons, 29
Photons, 29
Polyurethane application, 49
Polyurethane foam, 59
Polyurethane foam church roof, 56
Pottery kilns, 95

Reflected radiation, 66
Refractory insulation, 95
Refrigeration, 39
Refrigerator cabinets, 47
Rigid facings, 59
Rigid foam, 45, 57
Rigid polyurethane, 46, 47
Rockets, 25

Sand pugging, 87
Santocel, 9, 12, 13, 17
Shock absorbent mountings, 90
Silver, 117
Sintered alumina, 113, 115
— beryllia, 117
Slag wool, 5, 7, 9
"Soft" furnishings, 90
Sound, 84
— absorption, 83
— insulation, 84
Sprayed foam insulation, 55–57
Stainless steel, 116
Steel framed panels, 53
Strength, 110, 112
Structure-borne sound, 85
Summer, 62

Superinsulation, 17, 19, 20, 29
Superconductivity, 27
Surface resistance, 68
Surface topography, 29

Thermal conductivity, pure
   metals, 1
— —, alloys, 1
— —, dielectric crystals, 2
— —, gases, 2, 8, 18
— —, insulants, 2
Thermal radiation, 10
Thermal/sound problems, 86
Titanium alloy, 116
Top-hat kiln, 96, 97, 101

Transmitted radiation, 66
Truck-type kiln, 97

Ultra violet, 64, 65
U value, 67
— of glazing, 71

Vacuum insulation, 18, 22, 29

Water vapour, 38
Wilson report, 83
Window, 62
Window frame thermal bridging,
   76
Winter, 62